COMMON SENSE FOR THE 21st CENTURY

Only nonviolent rebellion can now stop
climate breakdown and social collapse

Roger Hallam

Editor: Arthur Ford

First published in 2019 in the UK by:

Common Sense for the 21st Century
Wern Dolau
Golden Grove
Carmarthenshire
SA32 8NE
UK

www.rogerhallam.com

ISBN: 978-1-5272-4674-4

Printed and bound in the UK by CPI Group (UK) Ltd, Croydon, CRO 4YY. Printed on 100% recycled paper using non-toxic inks. PDF version hosted on a green server.

Distributed by Turnaround Publisher Services, www.turnaround-uk.com.

A CIP catalogue record for this book is available from the British Library.

Contents

Acknowledgements
Matt Black, Ninja Tune, Peter Quicke, publishers Audley Burnett and Lindy Burleigh, Bill Godber and Claire Thompson of Turnaround Publisher Services, Paul Gilding, Katie Bond, Benjamin Tinholt, Pelle Hyek, and Maggie Taylor-Sanders of PCCS Books, other draft commentators, anonymous crowdfund major donor and all our crowdfund donors.

Cover art
Anonymous painting photographed during Extinction Rebellion action in Marble Arch, April 2019.

For press enquiries and feedback:
roger.hallam.uk@gmail.com

Please support our crowdfund to produce, distribute and translate this booklet:
https://fundrazr.com/commonsense

Foreword

I was there. For the past 20 years. Climate activism. It didn't work.

We protested in our hundreds of millions – it didn't work.
We raised billions to reach people and politicians – it didn't work.
We signed countless online petitions – they didn't work.
We looked to the United Nations to resolve the crisis – it didn't work.
We trusted progressive politicians and their reforms – it didn't work.
Al Gore had a big concert and a PR campaign – it didn't work.
Countless NGOs did their best – it didn't work.

I worked with green NGOs' campaigns, champagne environmentalists and lying politicians. I was wasting my time. I had a clue back in 2007 that there might be a fundamental flaw with the reformist approach. The problem of the political influence of the industrial billionaires like the Koch Brothers and other fossil fuel bosses.

It was another warm night in Los Angeles. I was invited to an exclusive event. A Tesla Roadster prototype was going to be shown to press and guests. We took turns to ride around the block. I watched in amazement as the Tesla was at the end of the street in the blink of an eye. With barely a sound. I felt like a caveman seeing a television. Incredible.

Whilst working on a campaign in 2007, I had an idea for how to save the world with a massive music event and a huge fundraising effort, led by celebrities, to fund the biggest PR campaign to save the environment. It would be a huge global event just like Live Aid. I sought out a prominent journalist at the event to tell about my scheme and bask in the glory of his praise. The journalist was Vijay Vaitheeswaran from The Economist, now a life member of The Council of Foreign Relations. After listening carefully to my concept, he smiled and laughed. I was stunned. Laughter? How could I be so wrong?

Vijay wasn't being cruel or unkind. He thought I was naïve. He said, 'The fossil fuel industry has enough money to outspend any public relations campaign you might try by 100 times.' He carried on talking, but I stopped listening. I had my answer. He was right. Crestfallen, I sheepishly nodded and walked away.

Over the next 12 years I watched as fossil fuel companies used their vast financial capital to buy political parties, tricked NGOs into fruitless schemes like the US Cap and Trade initiative, sowed doubt about climate science in the public discourse and slowed policy reform to a snail's pace. Their strategy worked. As climate scepticism continued to spread, climate change mitigation dropped down voters' priorities. With Putin and Trump in power, it was over. I quit climate campaigning.

Then I heard about Extinction Rebellion closing five bridges in London. Then they closed central London in April 2019. This was something different. I met Roger Hallam and he explained his ideas for a radical plan to create real change, this included:

> First – *Tell the truth*. Climate scepticism was pushed by the mainstream media which spread doubt on climate science like a plague, leaving inaction in its wake. The scale of the crisis is being played down and we must get the truth out.
>
> Second – *Nonviolent civil disobedience*. We need to get arrested, tens of thousands of us. More. No more protests or petitions. Instead, nonviolent civil disobedience, lots of it and on a large scale. Close down cities until the politicians take action. Or until the people do.
>
> Third – *Universalism*. In other words, the ecological and climate crisis should not be owned by any political ideology, culture, age or gender. As Roger Hallam says, we all have a stake in the future, we should all be allowed to engage in action.

I was convinced. We need a new approach to save ourselves and the planet. We need to spark a worldwide Rebellion.

We need some Common Sense for the 21st Century.

Anonymous climate activist

Introduction

It is time to grow up and see the world as it is. There are some things which are undeniably real, there are some things we cannot change, and one of those is the laws of physics. Ice melts when the temperature rises. Crops die in a drought. Trees burn in forest fires. Because these things are real, we can also be certain about what the future holds. We are now heading into a period of extreme ecological collapse.

Whether or not this leads to the extinction of the human species largely depends upon whether revolutionary changes happen within our societies in the next decade. This is not a matter of ideology, but of simple maths and physics.

The United Nations has estimated that we need to reduce carbon emissions by half within a decade to have a 50% chance of avoiding global catastrophe. Of course, this is likely to be an underestimation as recent science shows permafrost melting 90 years earlier than forecast and Himalayan glaciers melting twice as fast as expected. Feedbacks and locked-in heating will take us over 2°C even before we factor in additional temperature rises from human caused emissions over the next ten years.

In short, we are fucked – the only question is by how much and how soon? Do we accept this fate? I suggest we do not. Many self-respecting people who can overcome the human failing to disbelieve what they don't like, now accept what is obvious looking at the natural science. But they have yet to work through the political and social implications. This booklet sets out what is now obvious from a social scientific point of view: societies will not change with the necessary speed without rebellions and a revolutionary transformation of our societies and politics. This is not a matter of one's political party preferences. It is a matter of basic structural sociology. Institutions, like animal species, have limits to how fast they can change. To get rapid change they have to be replaced with new social systems of policy, practice and culture. It is a terrible and painful realisation, but it is time to accept our reality. Just because we don't like something does not mean it is not true.

All of this is common sense.

In 1776 Thomas Paine wrote a pamphlet titled 'Common Sense' to tell the American population what they privately knew but did not have the courage to declare openly: that they needed to declare independence from the British Crown. It was read by 10% of the population and is credited with transforming the resolve of many Americans to strike out into the political unknown.

Likewise, my proposition in this booklet is a Common Sense for the 21st Century; I declare what we already know – things cannot carry on as they are. Only a revolution of society and the state – a similar turn that Paine urged the Americans to take into the political unknown – can save us now.

This is the first step in transformation: accepting the truth as it is. Climate and ecological breakdown will kill us all in the near term unless we act as if the truth is real.

In this booklet, I explain the social implications – that the reformist political culture of both left and right in neoliberal society is now not fit for purpose. To put it bluntly, NGOs, political parties and movements which have brought us through the last 30 years of abject failure – a 60% rise in global CO_2 emissions since 1990 – are now the biggest block to transformation.

They offer gradualist solutions which they claim will work. It is time to admit that this is false, and it is a lie. They therefore divert popular opinion and the public's attention and energy away from the task at hand: radical collective action against the political regime which is planning our collective suicide.

The paradigm shift is to move from the words to radical action, from lobbying to mass breaking of the law through nonviolent civil disobedience and from elitist exclusion to popular democratic mobilisation.

The proposal
We must adopt the most successful model for regime change shown by the social scientific research – the civil resistance model. This involves mass participation civil disobedience: tens and hundreds of thousands of people blocking the centres of cities to demand change.

There are many tactical options, but the main process is as follows:

- The people conduct mass mobilisation – thousands need to take part.
- They amass in a capital city where the elites in business, government and the media are located.
- They break the law – they have taken that irrevocable step. Examples include blocking the roads and transport systems.
- They maintain a strictly nonviolent discipline even, and especially, under conditions of state repression.
- They focus on the government, not intermediate targets – government is the institution that makes the rules of society and has the monopoly of coercion to enforce them.
- They continue their action day after day – one-day actions, however big, rarely impose the necessary economic cost to bring the authorities to the table.
- The actions can have a fun atmosphere – most people respond to what is cultural and celebratory rather than political and solemn. Such an atmosphere also reduces tension, and keeps people calm and positive.

After one or two weeks following this plan, historical records show that a regime is highly likely to collapse or is forced to enact major structural change. This is due to well-established dynamics of nonviolent political struggle.

The authorities are presented with an impossible dilemma. On the one hand they can allow the daily occupation of city streets to continue. This will only encourage greater participation and undermine their authority. On the other hand, if they opt to repress the protestors, they risk a backfiring effect. This is where more people come onto the street in response to the sacrifices of those the authorities have taken off the street. In situations of intense political drama people forget their fear and decide to stand by those who are sacrificing themselves for the common good.

The only way out is for negotiations to happen. Only then will a structural opportunity open up for the emergency transformation of the economy that we need. Of course, this proposal is not certain to work but is substantially possible. What is certain, however, is that reformist campaigning and lobbying will totally fail as it has for decades. The structural change we now objectively need has to happen too fast for any conventional strategy.

I propose, however, that a material analysis of power is necessary but insufficient to maximise the chances of revolutionary success. Social relations are built upon cultural rules as much as material interests. It is therefore important to attend to the symbolic and interpersonal as much as the material. Mass action cannot just be nonviolent in a physical sense but must also involve active respect towards the public and the opposition, regardless of their repressive responses.

This not only undermines the regime's ability to 'other' you but also makes it much easier for it to save face when it comes to negotiations. Successful mass actions, then, have to include three aspects to optimise the chances of success: mass disruption, mass sacrifice, and respectfulness.

The mobilisation has to be built by creating alliances between networks. Most political networks are controlled by gatekeeper elites which have little or no interest in moving from a reformist to a revolutionary paradigm (even if they claim to believe in such a view). Nor will they wish to combine in a mass mobilisation in which they cannot control their particular ethical rules or identity orientations.

Therefore, it is necessary to use direct action to highlight their hypocrisy and to appeal directly to members of these groups to join in a universalist struggle – that is, to save all human beings and fulfil our transcultural duty to create a world safe for our children. This will involve appealing to people in a diversity of political, cultural and religious groups. The appeal will be successful if it addresses both the universalist threat and the opportunities provided through unity of action.

This then leads on to the thorny topic of framing – the way in which we communicate the message. Only when the message is put into a culturally neutral language can a mass movement be built. Some groups which have a high level of self-identity will find working with others problematic so it will be necessary to create separate mobilisations which will combine together into a 'movement of movements' rebellion. However, most ordinary people will be attracted to getting involved to the extent to which they are personally welcomed into the movement. This requires close attention to the design of all meetings. These methods and attention to detail may seem inconsequential, but they are of vital importance in order to create an effective inclusivity.

Finally, there needs to be a post-revolutionary plan otherwise chaos will ensue. The plan I outline is for a national Citizens' Assembly to take over the sovereign role from a corrupted parliamentary system. Parliament would remain, but in an advisory role to this assembly of ordinary people, randomly selected from all around the country who will deliberate on the central question of our contemporary national life – how do we avoid extinction?

They will decide what new structures and policies are necessary to maximise the chances of achieving our collective desire to live, now that the odds are stacked against us. We need to start acting now. We may need to act before governments come around. A transition movement has already started. This needs to be massively expanded and integrated with the Rebellion.

As this booklet goes to press, conflicting news comes that $1.9 trillion is being invested in gas and coal, whilst electricity generated from solar panels and wind turbines are almost on the verge of being cheaper than fossil fuels globally, and already are in many countries.

There's no time to lose. Let's get to it.

It's going to be quite a show.

Why a Rebellion?

Earth, we have a problem to solve

Societies around the world did not allow the current ecological collapse. Governments did. Since the 1990s, a false narrative was promoted around the world that individuals should take responsibility for their 'carbon footprint'. Or that 'it's the corporations', the fossil fuel and other polluting industries that are to blame. Yet governments are the only institutions with the power, and the responsibility, to protect us from harm. But they haven't used that power.

In the UK and around the globe, people have inherited a governmental system and a civil society community of environmental NGOs unable to address the threat we now face to the continued existence of humankind. Government is something created by society to protect us from such threats. Yet it has failed.

We need to rescue the concept of revolution from a rigid left-wing political ideology and reconnect with a more open, popular 19th century tradition which demands we've had enough of corruption and the gross abuse of power. The problem resides with the state and its capture by the corporate business class.

There is no avoiding the following analysis: that the world's present political systems have facilitated a 60% increase in global emissions since the beginning of the crisis in 1990 and have no ability to stop a continued rise in CO_2, let alone create the political will to massively reduce levels (40% in the next ten years according to the UN October 2018 report[1]).

This leads us to the grave conclusion that the probability of organising a political revolution to remove the corrupt political class has a higher chance (if small/indeterminate) of succeeding than the chance that the political class will respond effectively to the climate crisis (zero chance, as evidenced by the last 30 years). The penny has finally dropped – the corrupt system is going to kill us all unless we rise up. This then is the central meta-strategic point of this booklet.

The failure of reformism

Whilst there is no one clear moment we can point to, there are a number of factors that contributed to the disastrous delay in addressing climate change, and nearly all of them are linked to the emerging neoliberal political economy of the 1980s, the financial power and political influence of the fossil fuel industry and flawed thinking about how to solve the crisis within the existing political systems.

One example of a proposal was the 700-page Stern report published in 2007, written by Nicolas Stern and his colleagues at the London School of Economics Grantham Institute.[2] One of the central ideas proposed by Stern was called the 'discount rate' of mitigation costs. This idea suggested that it was possible to act on climate change further in the future as technology developed and renewables became cheaper.

Stern would later adjust the discount (or savings) rate he originally suggested but the flaw at the heart of the discount rate was that it assumed perfect political and economic conditions. For example, it assumed the political willingness to invest in new technology in addition to providing large subsidies to renewable energy, green transport and less polluting industries. This did not happen because the fossil fuel industry created social and political discord by funding anti-climate change science groups which manufactured doubt about human influence on the climate.

The same naïve assumption of perfect conditions can be found in the structure of the United Nations Framework Convention on Climate Change (UNFCCC) which uses a consensus decision-making system allowing small groups of countries to veto any decision. Additionally, the UNFCCC never became an effective governance system to make commitments legally binding. This meant a positive outcome relied on rational economic and political decision-making processes in an irrational world.

The UNFCCC assumed that world leaders would support and ratify the outcome from the UN Paris Agreement of 2016 as well as the beneficial economic conditions needed to finance the transition from fossil fuels. Yet the 9/11 attacks, subsequent wars and the financial crash had pushed the growing climate crisis off the political agenda. The result was the weak Paris voluntary agreement that is being ignored by the US, Russia and many other governments.

The EU cap and trade system: more flawed political economy
More utopian thinking can be found in the foundation of 'step by step' gradual reform processes like the EU cap and trade system whereby high carbon industries were forced to pay for emissions. However, a major flaw in the policy made the system ineffective. By allowing all high emitters a free carbon allowance to start with, it gave many corporations time to move their production facilities to China without penalty. Others simply paid the fines.

All of these examples relied on the good will of politicians, the good will of corporations, pure rationalism and perfect national and international conditions. None of these transpired and the reformist process had no Plan B. Meanwhile, the political systems that are supposed to protect us became increasingly 'bought' by the fossil fuel lobby whilst simultaneously exploiting concerns about the environment to obtain votes from centrist voters.

Needing to keep voters, and having their hands tied by powerful vested interests, governments worldwide are seemingly unable to make the necessary changes. That's why a Rebellion is necessary.

Accepting the truth is the first step

What is the 'truth'?

We have to start with what is true. One of the main problems we have experienced with climate change and environmental activism is that people rarely seem to talk about empirical reality (i.e. the latest science) and thus aren't even aware of how desperate the situation actually is. We have even 'chosen' to rubbish the science and the scientists because they are not telling us things we want to hear, so it's easier for us to believe they are wrong.

This is a form of denial that many do not realise they have fallen victim to.

Campaigners often believe they know the magnitude of the problem we face. However, 'knowing' is not a binary concept when it comes to grief-inducing catastrophic information. You think you know it but then you realise you haven't processed it emotionally. You think you have processed it emotionally but then you realise that you haven't done it sufficiently. This leads to a form of 'unconscious denial' which in turn leads to a form of personal protectionism.

People live with their unconscious denial because to accept the truth and act like it was real might endanger their relationships with family, friends and co-workers (see Professor Dan Kahan's research from Yale on science communication[3]). The issue of climate change became a taboo issue which led to a default of inaction from the general public beyond basic recycling activities.

A quick update on the latest science

The science matters. The IPCC reported in October 2018[4] that we have to reduce carbon emissions by 40% in the next 12 years to have a 50%

chance of avoiding 'catastrophe'. And yet in 2018 emissions went up from an increase of 1.6% in 2017 to an increase of 2.7%. Carbon dioxide levels went up by 3.5 parts per million (ppm) in the past year to reach 415 ppm. We are now only ten years away from 450 ppm, the level equivalent to 2°C average temperature rise.

Let's be frank about what 'catastrophe' actually means in this context. We are looking here at the slow and agonising suffering and death of billions of people.

A moral analysis might go like this: one recent scientific opinion[5] stated that at 5°C above the pre-industrial mean temperature, we are looking at an ecological system capable of sustaining just one billion people. That means 6–7 billion people will have died within the next generation or two. Even if this figure is wrong by 90%, that means 600 million people face starvation and death in the next 40 years. This is 12 times worse than the death toll (civilians and soldiers) of World War Two and many times the death toll of every genocide known to history. It is 12 times worse than the horror of Nazism and Fascism in the 20th century.

This is what our genocidal governments around the world are willingly allowing to happen. The word 'genocide' might seem out of context here. The word is often associated with ethnic cleansing or major atrocities like the Holocaust. However, the Merriam-Webster dictionary definition reads 'the deliberate and systematic destruction of a racial, political, or cultural group'.

With many governments knowing the impact of climate change but continuing to support the fossil fuel industry, the result will be the destruction of many nations, species and cultures. There is no greater crime. Let us bear this reality continually in mind as we address the question of the necessary societal and political response.

What is to be done?

Successful political campaigns rely on a number of factors, some we can control, others we cannot. Successful radical political action also relies upon the application of a specific method – rigorous and detailed empirical investigation of the field in which a political conflict is going to take place. As opposed to an idealist, 'perfect conditions' or 'how we would like it to be' approach, we should design our collective action for imperfect conditions, for political instability and, most importantly, for counter-revolutionary tactics from the fossil fuel industry. It involves an analysis of all the players in the specific time and space where the confrontation is going to take place.

Design principles

A first practical design principle is to choose the time and place where the conflict will occur rather than reacting to the opponent. This gives us time to 'map the territory' and choose a terrain where we can maximise our chances of mobilisation and thus of success. This enables us to concentrate all our resources on a single time and place. Therefore, we must know the actions and intentions of the main players – our activists, our supporters, the opponents, and their supporters.

A critical group here are the opponents, who can be broken down into the elites/management and the various supporting pillars and groups who follow their orders: courts, police, security people etc. In nonviolence theory it is important to notice that the interests of the elites and those that carry out their orders are not equivalent, and indeed in contexts which provide structural opportunities for revolutionary change, their interests can significantly diverge.

We need to focus on the specific blocks to mobilisation and campaign success and then think about a range of micro designs to maximise the chances of reducing these blocks. For instance, I designed London's first organised rent strike by working out that the key block was not that people were not pissed off with their rent levels. They were. The problem was that tenants would not go on strike because they were not confident others would do likewise.

I developed a 'conditional commitment' routine for canvassers. Would you go on strike if a critical mass would act likewise? This worked and led to a hall going on rent strike. The fear factor of eviction disappeared and many more joined the strike, leading to an agreed rent reduction.

Avoiding ideological dogmas and learning from history

This is a world away from how many people think about 'how to win'. A key problem is that they take a historically similar situation which did not work and then project this outcome upon a present context as a way of saying 'your action plan won't work either'. Instead we need to look in detail at how the present situation and proposed plan are similar or different from what has happened before. Small differences and the introduction of new design elements can massively change outcomes in a complex social system. This problem is made worse when ideological dogmas are imposed on scientific facts. We need to look at empirical feedback objectively.

We cannot see clearly because we are closed to various outcomes. A simple example is that, 'the police are nasty for ideological reasons x and y' and therefore it is difficult to accept that the police can be cooperative in a particular set of circumstances. This then prevents us from taking advantage of new data and therefore new possibilities.

It is important to note that the 'best design' is not one which will work but one that maximises the chances of success relative to the other options on the table. Therefore, it is possible that the plan has only a small chance of success, but it has to be compared with other options that might actually solve the crisis, not theoretical options which are constructed within the current corrupt political and economic ideologies and thus far have failed.

This is the practical choice we face. We have to get into a position where we get to 'roll the dice' which might lead to structural political change. We have to take the chance even if the odds are stacked against us. This outside chance is the only one we have for effecting the necessary changes to save our communities, nations and the environment. Small though the chance might be, we have to 'roll the dice' because 'it might just work'. This is the logic of pursing rebellion in a complex social system.

Designing a pathway to success

With this in mind, it is worth introducing a method used in other pre-organised civil resistance events. This is 'reverse engineering'. Instead of working from the here and now to success, we first work out what success looks like and then work back to how it would be created through time from then to now. This has the significant psychological benefit of looking back from the mountain top rather than looking up.

We have succeeded and just have to work out how it was created. It therefore encourages us to take greater chances and risks without which success is evidently not going to be even possible. The issue is not that the risk might lead to failure but that by not taking the risk, failure is guaranteed. All options are now dangerous and risky given the unavoidable existential crisis in which we now find ourselves.

Reformism vs political revolution

Reformism to revolution

The political culture of Western democracies has changed from a reformist to a revolutionary structure. It is no longer possible to save our society through small incremental steps, as happens with reform. Mass political disruption is now required. This is a sociological observation rather than an ideological assertion – that is, it is based upon evidence. The evidence can be found in the devastating failures of the reformist political class to correctly predict the outcome of a whole series of political contests in recent years.

Arguably this failure started with the Arab Spring and the assumption that the Arab people would never rise up against dictatorship, still less, win. Then we had the meteoric rise of new left-wing parties in Greece and Spain – from effectively zero support to 30% to 40% support in two years. Then there was the 'intense embarrassment' of the Corbyn win and the ability of the Sanders campaign to mobilise two million people for political revolution in the US. We should also mention the darker side – the total 'surprise' of Trump.

The approach of reformism (and I am not making any ultimate moral judgement here) is that progress is maximised by making small demands and small 'asks' of your supporters. The logic is that this is more 'credible' because it makes some progress rather than none. The argument then is that making radical or even revolutionary demands is not credible and therefore leads nowhere and thus is ineffective compared with the reformist approach.

This is true in a reformist political context – this is where the common view is that society is mostly stable and the problems that exist can be sorted out by gradualist campaigns that make small demands, issue by issue.

The problem is that, sociologically speaking, not all contexts display signs of a reformist political culture. Some have a revolutionary political culture. Such a phenomenon is evidenced by mass disillusionment and distrust towards the political class and a high level of social repression. People conform but don't want to. This then explains the error of conventional analysts, such as the recent failures to predict the results of public votes. On the surface things look like a reformist political context – nothing revolutionary is happening because of the repression and so they presume it's business as usual.[6]

What happens then is that when this repression finds an outlet there is a non-linear political event, an event that doesn't follow a logical straight line. Politicians like Corbyn offer radical programmes which we are told are not 'credible', but the possibility for change provides the outlet and pathway, and people are drawn to the new opening in vast numbers.

Extinction Rebellion was set up in April 2018 to 'Tell the truth and act as if that truth is real' on the climate and ecological emergency. Again, this was not a credible approach, and the same thing happened, but in the political campaigning and social movement sphere, rather than the party-political sphere. The structural analysis is the same. Extinction Rebellion said what a lot of people were thinking and proposed a pathway to action, 'We are facing extinction due to the ecological crisis and that we should take radical collective action, which means engaging in a rebellion against the government.'

A conventional view – e.g. the one I got from the chief executive of Greenpeace who I met two years ago – was that such an approach lacked 'credibility' and therefore, would fail. This view is encased in the reformist space which has dominated politics from 1989 until the 2007 financial crisis.

Times have changed.

'Tell the truth – then act as if that truth is real'

The statement 'Tell the truth and act as if that truth is real' is an extreme violation of the reformist paradigm. For reformism you only tell the truth to the extent that you think people can cope with it and you only act on it to the extent that you think you can win (in a gradualist way).

This is how reformism ends up in a morally and spiritually bad place – lying and holding back actions which are now justified. So what is the revolutionary alternative?

A proposal for Rebellion

The core proposal
There are then two opposing broad logics going forward. There is the reformist logic; to engage and become 'credible' and settle for a gradualist progression. And there is the expansion of the revolutionary logic on the basis of the successes experienced so far. I want to argue that it is now necessary to pursue the latter.

In the event of the unwillingness of the government and the elites to respond even minimally to demands for structural change, we must draw the conclusion that, due to the dire crisis we face, only a change in the political system itself can lead to our demands being rapidly enacted.

This could take the following form. We issue a public statement that if, by a specific date, the central government has not responded to our demands and started to enact credible measures to respond to the existential climate and ecological emergency, we call for mass nonviolent civil disobedience.

The demand
Given the complete moral failure of the government to respond, the demand should be – 'The current government hands power to an administration which will call a national climate and ecological emergency and immediately enact measures to deal with the climate and ecological crisis.'

What next? The people decide
It is one thing to propose a rebellion or revolution, but it is another to work out what happens next. There has to be a plan for a credible and attractive alternative arrangement in place. This is a National Citizens' Assembly selected by sortition to work out the programme of measures

to deal with the crisis. Sortition involves selecting the members of the assembly randomly from the whole population and uses quota sampling to ensure that it is representative of the demographic composition of the country. This proposal then is both concrete and democratic.

The National Citizens' Assembly will become the new governing body of the UK and will deal with the climate crisis. It will make decisions on the following:

- Legislation to transform the economy and society to respond to the existential climate and ecological emergency.
- Other social legislation which follows the will of the assembly rather than the former political class.
- Draw up a new constitutional settlement which creates a genuine participatory democracy fit for the 21st century.

The revolutionary context

There is more than one revolutionary context. Some are more obvious than others, but all are difficult to assess because they involve an awareness of what is largely hidden – people's private desire to be rid of the failed regime. Or potentially, and more importantly, the number of people who are indifferent to whether the political system is removed or not.

Some things are important to remember here. We need only a few hundred thousand people to actively break the law and/or support such activities to put us in the ballpark of structural change. We can see that five days of insurrection by 200,000 people in France produced a back down by the government. We should not make the mistake of thinking 'the people have to rise' in the sense of the majority of the population. We need a few to rise up and most of the rest of the population to be willing to 'give it a go'.

Structural weakness within elites

The great structural weakness of any elite is that the seeds of its destruction are created by its very success at domination. Success creates separation from real life and a bubble of self-reinforcing orientations – namely that everyone thinks our system is fine. Secondly, elites start with a virtue ethics orientation, which places emphasis on the role of an individual's character and virtue in moral philosophy rather than an individual's duty or actions in a pro-social and pro-environmental way.

They believe in ethical values, but they end up only believing in two things – money and power. This process of degeneration is now well advanced.

From the other end of the spectrum, the belief in the political system from those on both the left and the right has withered since the financial crisis, as evidenced by the rapid growth of radical political groups which use anti-system rhetoric. There is growing rage at the injustice of extreme inequality and the unaccountable global elites and now we have the emergence into public consciousness that not only have these people been robbing us for 30 years, but they are now going to take us to our deaths.

Bring down the government you say?

I would argue that the slogan – 'Bring down the government' (or similar) – has an incredible (and therefore actually a very credible) ring about it. Its attraction is that it fully releases the social repression in the most clear and explicit way – we want to get rid of it. It is simple, concrete and dramatic.

Question: 'What do we want?'

Answer: 'To bring down the government.'

You can imagine the elites and the establishment saying, 'How ridiculous!' whilst ordinary people say, 'Well, it can't go on as it is'. Rebellion is about giving people permission to say what they really think. It creates the connective tissue for solidarity that many don't realise exists.

This releases enormous political energy and imagination. This is what has powered the rapid explosion of support for Extinction Rebellion. Rebellion is ridiculous but for that very reason it is appealing. It's transgressive and people want to break the existing rules and love to see that now some set-up is finally going for it. There is a deep psychological attraction to going into the unknown in a world where we are only offered the option to 'Put up with it and fit in'. There is much to indicate that in crises people seek meaning in what is happening, as well as material security. This is supported by the evidence of the last nine months.

The materialist analysts on both the conventional left and right have missed this because they both accept the reductive dogma of neoliberalism that money and consumerism dominate our lives. A broad historical analysis shows this is plainly wrong.

Revolutions often fail

On the other side of the equation there is a big reason why revolutions do not work – because there is no credible pathway for order to come out of inevitable chaos. If there is no plan, then a political vacuum appears which leads to the escalation of revolutionary violence and/or people falling back into the arms of the old elites.

This is not a reactionary consideration. Permanent revolution is hell, not heaven, as the record of the 20th century shows. People want change but, contrary to political enthusiasts of all shades, they do not want politics to dominate their lives. There are more important things to consider: family relations, creating beauty, gossiping, having a laugh, making a living – the constants of all human societies.

The political revolution that is proposed here has to find a tricky balance between a joyful exuberant celebration of collective human agency – we will decide for ourselves and such like – and creating a brake on utopian excess. The fundamentals of life remain the same: we still have to face death, we have to master ourselves and learn to relate effectively with others and so on. Politics is a significant part of life, but it is not all of life. This is where we can move on from politics as domination which characterised the 20th century.

The key element in persuading people that the whole thing will not end in tears is that we have an answer to the most fundamental question of politics – 'Who decides?' The main reason why revolutionary episodes have failed miserably over the past 30 years is that revolutionaries have either no answer to this question (e.g. the short-lived Occupy Movement) or fall back on the representative parliamentary answer which has been shown to be irremediably corruptible in the context of the dominance of a global capitalist system – that is, big money trumps political independence every time.

We see the latter in Egypt and Ukraine where an amazing show of people power resulted in a return to the old regimes, because there was no plan for the day after the revolution.

The Action Plan

The starting principle of the plan is simple; all resources – trained activists ready to be arrested, tents to sleep in should the event go beyond a day and artists and musicians to create a positive atmosphere – should be applied to a single point in time and space. This maximises the chances of increasing our political power up to the binary tipping point where on the one side nothing happens and, on the other side everything happens. From the government refusing to agree to serious policy change to the point where it calls for negotiations with a nonviolent rebellion.

We won't agree to any compromise that allows the incompetent and corrupted political class to remain in power. They are forced to choose between either agreeing to our demands or repressing the actions and protests with the risk that this inspires more people to join the Rebellion and take to the streets.

Our aim then is to gather enough resources to reach this critical point where we force the hands of the politicians to make the choice; agree to give up power or repress us.

Legitimacy and credibility – critical elements to effect change
Two key issues need considering in relation to attempting to force political change: political legitimacy and mass scale disruptive actions. In response to acts of disruption the standard universal response is 'why you?' Why does any movement have the right to tell society what to do? I believe we have a credible answer: if we don't change quickly, we will soon be dead. This realisation helps concentrate the mind on serious changes based on science rather than options that reformists and corrupt governments would like to have.

However, there is a second answer which is just as powerful – a democratic assembly of ordinary people which has decided we need emergency action in the absence of any credible response from the political class. The National Citizens' Assembly will have more democratic legitimacy than the elite and their corrupt politicians.

The broad aim is to do two things:

- Continue to organise and inspire nonviolent civil disobedience around the world until serious action from the political class takes place.
- Organise professionally constructed nationwide Citizens' Assemblies, selected by sortition, to give a judgement on the need to declare and act upon a climate emergency.

The science demands action, citizens are calling for serious policy, but the government has no more legitimacy given its failure to act meaningfully. Governments have been criminally inactive in the face of warnings from scientists, and the only way we can have a chance of saving our children and enacting the will of ordinary people around the country is to replace the political class.

Citizens' Assemblies are not a reformist process

It has been suggested that holding a Citizens' Assembly without government participation will result in a talking shop that will be ignored. This has definitely been the case within a reformist paradigm as we have seen with the United Nations Framework Convention on Climate Change where the United Nations was tasked with bringing the entire world to reach a consensus on climate change action. However, with ecological collapse at our doorstep, Citizens' Assembly decisions and policies will be supported with mass civil disobedience in order for any policy they agree on to be implemented.

As discussed, the reformist and revolutionary logics are at odds with each other. What makes logical sense as a theory of change, reformism, is totally nonsensical to revolutionary theories. In the revolutionary context of the climate emergency, the holding of a Citizens' Assembly on this crisis becomes a revolutionary act.

Citizens' Assemblies reveal their dramatic political power through the profound effect on ordinary people of seeing people like them (as opposed to activists) declaring a climate emergency to the world, and making it clear how accepting this reality has led them to support a transformation of the economy. This declaration gives popular legitimacy to the whole Rebellion project enabling it to attract mass public support and acceptance from the undecided. Having a cross-section of the public deliberating on the central crisis of our time, rather than politicians, is the key to the assembly having a major impact on our social and political culture.

The central point is that we need to push everything we have into this one point in time and space and this includes the Citizens' Assembly process. (Read more about this in the chapter on post-Rebellion, pp. 62–66.)

Major civil resistance and build-up actions

The historical record shows that successful civil resistance 'episodes' last between three to six months. It is not a matter then of 'take it slow and safe'. This is even more the case now that we face an existential emergency.

The build-up period to a major act of civil resistance, the culmination of smaller actions, needs to be planned carefully. Too many build-up actions and people will get exhausted, too few and people will not have the confidence. The aim of these movement-building actions is to prepare the frontline Affinity Groups.* Three points need to be made:

1. It is essential that build-up actions and event dates are planned in advance so that people nationally and internationally can put them in their diaries.
2. Similarly, people need to be told about the start date of the major civil resistance event. It's 'all-hands-on-deck' and people need to mass at the pre-agreed points and stay there 'for as long as it takes'. People can then take time off work, tell their families and prepare for arrest and prison. We know from Extinction Rebellion actions that after people have decided to break the law, the biggest block to

---

* Affinity groups. These are groups of 8–12 people who work together and support each other on the day of the action and preparing for it. They include one or two support people who do not get arrested and will stay in touch with the main working group throughout. Affinity groups meet before actions and agree on their joint activity.

mobilisation is not giving people enough time to ensure they can attend the main dates.

3. It would be beneficial to the Rebellion for people to be in prison before the major civil resistance event to create national publicity. The best way of potentially doing this is for people to do repeated acts of peaceful civil disobedience and then read out statements as soon as they enter court, ignoring the judge and court staff. In a loud voice they might say 'I am duty bound to inform this court that in bringing me here it is complicit in the "greatest crime of all" namely, the destruction of our planet and children due to the corrupt inaction of the governing regime whose will you have chosen to administer. I will not abide by this court's rules and will now proceed to explain the existential threat facing all life, our families, communities and nation ...' and then start a long speech on the ecological crisis.

   This will likely result in the arrestee being in contempt of court and placed in remand or given a prison sentence. It will be a dilemma for the authorities (depending on the regime) as to how long the remand or sentence would be. If the period of imprisonment is short, then people will be out soon and can continue peaceful civil disobedience. If the sentence is long, it will create a national media drama which will feed into overall rebellion.

The dramatic potential of mass participation civil disobedience vs small-scale, high-risk direct action

For 30 years – since the demise of the 1980s peace movement – direct-action activists in the UK have been faced with a limited number of options due to their small numbers. Open actions, with the authorities knowing in advance, get stopped due to their low numbers compared to the police strength. Alternatively, activists organise secretly in order to reach their targets without the police or authorities being aware of their plans.

In the latter case, the stress of often getting discovered in advance is always present, which can lead to a closed and paranoid culture. An additional problem of secret direct-action planning is that fewer people know about them, and thus when structural conditions create opportunities for mass mobilisation, as in the present context, activists do not take advantage of them. Established groups tend to become attached to the ways in which they have always done things and find it difficult to change.

The Rising Up! network* was set up to question this conservatism – the unwillingness of activists to look at new strategies and tactics. The hope was that by trying tactics which 'will not work' according to conventional activist wisdom, we would find something that will. A key discovery here is that mass civil disobedience works better than any other strategy. This is no accident as the main social scientific research on the subject ('Why civil resistance works'[7]) shows that mass participation civil disobedience is the most successful strategy in bringing down regimes.

Whilst they seemed like a good idea at the time given the unwillingness of the media to cover environmental issues, small-scale direct actions (for example by NGOs like Greenpeace) have had minimal to no meaningful impact on government policy. This poses a challenge to many traditional direct-action activists who believe that high-risk, media-stunt actions can be as or more effective than the general public or 'activism novices' sitting in roads in their thousands.

There is often an implicit elitism in the direct-action environmental movements – an unwillingness to engage with the public directly and organise them to break the law on a large scale. Dramatic actions can be highly effective but only once thousands of people are already involved in civil disobedience. In such cases they can inspire more people to step up and break the law.

My argument then is that radical change is primarily a numbers game. Ten thousand people breaking the law has historically had more impact than small-scale, high-risk activism. This was demonstrated by Extinction Rebellion when thousands of 'first time' activists blocked the bridges in November 2018 and London streets during April 2019. These were the biggest civil disobedience actions for decades and were covered globally by the media. Their impact continues to have major inspirational effects around the world, but it required no high risks and was done peacefully. The key challenge then is to reduce the barriers to participation in such mass mobilisations.

---

* Extinction Rebellion arose in part out of the Rising Up! campaign group, which in turn arose out of the Occupy Movement, among others.

The importance of openness

Being able to mobilise large numbers opens up a third possibility not available in the traditional context of activism-as-usual. It means that lots of people can break the law (in a minor way) even if the police know about it. This allows campaigners to be completely open about the events. This is a game changer because it means thousands of people can learn about a proposed act of mass civil disobedience, which makes it more effective.

It is important to understand why this is important and that it has no bearing on the moral debate about whether the police are good or bad. That is a separate question and one on which we don't need to have any collective position. The key question is how can we make direct action work to the extent that it creates openings for the radical structural changes everyone wants?

Proactive and respectful police engagement

A proactive approach to the police is an effective way of enabling mass civil disobedience in the present context. This means meeting police as soon as they arrive on the scene and saying two things clearly: 'This is a nonviolent peaceful action' and 'We respect that you have to do your job here'. We have repeated evidence that this calms down police officers thus opening the way to subsequent civil interactions.

The Extinction Rebellion actions have consistently treated the police in a polite way when we are arrested and at the police stations, engaging in small talk and quite often in political discussions and other topics where activists might have affinity (inequality, unfair pay). If police initially stonewall activists, they can become more open by a willingness to engage with and listen to them.

This engagement can start before an action. Often a face-to-face meeting with police is effective as they are able to understand that the people, they are dealing with are reasonable and communicative.

Depending on the country or region, the police might not primarily be concerned with either upholding the laws we break nor being aggressive as long as activists are civil and open with them. This can be done by building up some informal protocols with the police and action liaison officers before actions happen. This produces something both sides want – predictability.

Crucially, this enabled Extinction Rebellion to predict that they could block five bridges by splitting into five groups at the same time. It also enabled us to predict that as long as everyone maintained nonviolent discipline, the police would behave in a civil way and arrest people calmly. Both these predictions proved correct and they enabled us to create this critically successful action. Similar predictions proved to be correct during the April 2019 rebellion.

The key to success was the trust that was built up between us and the police through having regular meetings. The police were assured that we would act in the way we told them in advance (as we have done on each occasion). Thus, from a risk analysis point of view, they did not prepare to overreact (e.g. have thousands of police in expectation of violent disruption). Secondly, it was in their interest to follow through on their stated intention to be civil to us so that we would continue to meet and tell them about future actions, something they are very keen for us to do.

When individual criminal actions are justified

The Ploughshares* direct actions movement is made up of people committed to peace and disarmament and who nonviolently, safely, openly and accountably disable a war machine or system so that it can no longer harm people.

Ploughshares activists are given training in safety and nonviolence and form groups for long-term support. Ploughshares is drawn from an enactment of the Biblical prophecies to 'beat swords into ploughshares' but is now no longer a Christian movement but one which embraces people from many different faiths or from none at all. The underlying appeal is the universal call to peace, to abolish all war and to find peaceful ways to resolve our conflicts.

It recognises the abuse of power that war always is and the deep immorality of threats to kill. Ploughshares actions started in 1980 in the USA and have taken place in many different countries with weapons as diverse as rifles, warships, missiles, submarines and aircraft being dismantled or damaged.

---

* For more about the history of the Ploughshares movement see Wikipedia.

In Britain a successful 'Seeds of Hope' Ploughshares action was one carried out by four women who did £1.5 million worth of damage to a British Aerospace Hawk jet.* The plane was prevented from being exported to Indonesia where it may have been used to continue the genocide being committed in East Timor. The women were acquitted in a landmark case at Liverpool Crown Court in July 1996 having argued that their act was justified in law as they were preventing British complicity in genocide.

A critical point here is that any act against property must have a clear ethical purpose and meaning rather than being a random act of damage. Beware of undercover officials who might try and cause trouble to delegitimise the action. If in doubt, ask who they are.

Incentivising first-time disobedience

Another vital advantage of maintaining civil relations with the police is that thousands of people can engage in their first positive act of direct action and come out of the experience reassured and willing to do more. This meant that the peaceful and inspiring blockage of the five bridges became a major movement-building event. If the police had overreacted, many first-time people (around 80% of participants) would have come away not wishing to 'take the risk' again. What was also interesting about the five bridges action was the timing of police arresting people (note this has little to do with actually breaking the law).

Establishing 'zones' for the blockades

In relation to the likelihood of arrests, blockades fall into one of three zones, which I explicitly discussed with liaison officers:

1. A zone where nothing critical is happening, e.g. just blocking a single road, in which case arresting people is not worth the time or hassle.

2. A critical zone when you are near a hospital or maybe have blocked too much traffic and police will opt for an orderly five-stage warning procedure and then begin arresting people.

---

* See 'Seeds of Hope' Ploughshares East Timor at https://wagingnonviolence.org/2015/10/seeds-of-hope-east-timor-ploughares-book/

3. Lastly, there is an emergency zone where a life-or-death ambulance needs to get through, or other urgent event, when the police will use whatever force is necessary to move protesters. In this case we agreed we would move, and they assured us they would not lie to us about it.

In the following examples, I outline two action designs which incorporate this new potential that has been opened up by having large numbers and thus an enabling police response as outlined above.

Blood of our children

This is a specific example of a number of similar actions. The design of this public event would put the police in the position of making it likely that they would arrest people in an orderly way. In this example people via social media and affinity groups would be recruited to gather outside government buildings.

They would bring red paint or other safe red liquid in buckets, meet at specific points, then walk to government buildings, stand in a line and throw the blood/red paint onto the ground and sit down as a symbol of the blood of the children who are likely to die.

The messaging and artistic design could be worked up, but the basic tactic puts police in a dilemma. We would actually tell them the plan. This would make it likely they would let it go ahead because if they try to intercept individuals beforehand the protesters would simply drop the open bucket on the ground causing a mess for arresting police.

Given there is no security risk, because police will trust that activists will pursue the pre-agreed actions, the likelihood is the protest would go ahead. They would rightly calculate it is easier to arrest everyone in an orderly way, in one go, outside the government building. In addition, they will have to arrest people because the action will have passed the 'criticality' line – throwing a lot of paint over the road is obviously criminal damage and you have the added 'bonus' of it happening in front of government buildings.

It is better for the police to manage an orderly and low-cost episode which is compatible with our interest in having a large number of people take part in a highly symbolic and dramatic act.

An action along these lines was carried out in March 2019 in front of Downing Street. As predicted, we were allowed to go to Downing Street. However, no arrests were made despite engaging in criminal damage. The 'political' response was to deny publicity of arrests made by pouring the 'blood of our children' on the ground to highlight the planned destruction of the next generation.

Blocking critical infrastructure

Another action design is an economic disruption. A primary candidate for such disruption would be to block the roads leading out of a major port, for example Dover, UK, through which most of our food imports enter the country. This could ideally involve several thousand people and take place over several days. Again, there would be several meetings with the police to assure them that it would be totally nonviolent, and we respect they will have to do their job and arrest people if they so choose.

The event would be publicly advertised and so the press and companies involved would know well in advance. Knowing in advance might enable people to change their plans, but this in itself would cause disruption and a possible overreaction.

Mass civil disobedience is best organised through open actions. Once we have a thousand or two thousand people taking part, it is very difficult to stop such significant disruption from happening. Again, the tactic is that the police will opt for orderly arrests rather than more aggressive attempts to stop the action – especially as we would have informed the police in advance of the peaceful nature of the planned civil disobedience.

The activists would organise shifts of affinity groups, ideally several hundred at a time, occupying the motorway. People would stay for a long period on the motorway until arrested.

Alternatively, this could be combined with a 'swarming' design, which is to move in and form a large, dense group, which could block motorways turn-offs/off-ramps and would continue until the threat of arrest became real, at which point the protest would be stopped. 'Swarming' is a fast-moving, swiftly adaptive action, like the behaviour of a swarm of bees moving from hive to hive. Such swarming could continue for several days.

This combination would result in national media reporting on a powerful event which involves many arrests and/or the effective blocking of a major port. Specific details and artistic contributions would come from the local groups. An initial successful event could then progress to another larger event which could involve several road or motorway blockages around other ports in the country.

The above two examples show what dramatic possibilities can be opened up by a proactive approach to the police and mass open organising. A new landscape opens up for radical participatory political action which has been missing for many years. By developing these options, we can create heated national debate and resultant mass attitude changes which are vital to legitimate the massive economic changes which are required.

The power of symbolic and sacrificial disruption

For 30 years the main emphasis for major direct actions has been the disruption of important economic infrastructure or other material structures. Direct action, as a way of creating political change, has been subject to a simplistic analysis that sees winning and losing in narrow material terms. There is a strong argument for this approach as confrontation, strikes, blockades, pickets, stoppages, economic threat and disruption can certainly bring opponents to the table – as shown by the long-term success of many labour strikes around the world.

However, we cannot assume that closing down a power station or a port is a simple pathway to success. And there are significant limitations to this approach. The Rebellion needs to change the cultural, psychological and indeed the spiritual. It needs to work with 'hearts and minds'.

Material structures exist within a larger social space subject to mass psychological dynamics. Direct-action design has to create desirable symbolic interruption – the meaning structures through which people interpret whether the disruption is justified. To work symbolically like this, the Rebellion needs to engage at many different cultural levels: with art, design, music, feeling and discussion. It needs to be 'human' and 'fun'!

Raising the economic costs for an opponent is highly provocative. If it's done in an abrupt or aggressive way, or at the beginning of an escalation, then onlookers are likely to side with the target as they may feel they've been unfairly impacted. It can also lead to the opposition hardening their response. In other words, there would be little change in the hearts and minds of people towards our cause; and the chance of gaining the 'moral high ground' can slip away.

Winning hearts and minds
There is a tendency for onlookers to see conventional direct-action tactics, e.g. lock-ons,* as 'something I have seen before', and therefore nothing of major concern. It is not emotionally engaging enough to create real interest. The response is, often subliminally, 'They're not like me and so it doesn't involve me'. The importance of changing hearts and minds cannot be overstated in terms of radical social change. What is important is not that we shut the plant down but more that we changed a lot of people's attitudes.

This point has not gone unnoticed in military theory. An occupying force can be powerful materially but will still lose against an insurgency if it does not actively engage with the civilian population (even at the risk of casualties) to win hearts and minds. The Iraq and Afghanistan conflicts are examples of where hearts and minds were not won and both wars have been a failure.

Therefore, material disruption needs to be designed to provoke a transformational national conversation and debate on what is going on with the climate emergency and ecological crisis. This animation of the public sphere is created through a national moral drama driven by the enactment of transgressive sacrificial and symbolic action on a large scale. Action which leads to arrest and prison.

This process is the key mechanism through which to create sympathy from supporters (leading to more recruitment) and grudging respect from critical onlookers – 'I don't like their tactics, but I give them this – they stand by their beliefs'. This is a good reason to stop using lock-ons and other physical equipment for blockades and to just use our bodies. The willingness to expose one's vulnerability in a fearless way is the key driver of an emotional response from observers. The message is 'I am just using my body and I am putting myself in harm's way because I feel so strongly'. Getting arrested through such actions is the classic sacrificial move. Having hundreds arrested in one day will be a major news story given the drama provided by such public sacrifice.

---

* Lock-ons are tactics used by individuals to make it harder for police to remove them from the scene of an action. This might be locking themselves to buildings with bike locks or using glue.

We must therefore avoid the simplistic idea that if we close down the country we will win. Disruption has to be combined with our willingness to show our vulnerability and to suffer. The disruption then simply sets the stage for the symbolism of fearless sacrifice. It is the sacrifice which brings about the social change not the disruption in itself. This is particularly the case when the young and the old take part as sacrificial action by both groups creates far greater public interest and sympathy than the 'usual suspects' – young men. This is the key reason why direct actions should be designed in an open way which are accessible to the two groups.

Creating a mass civil disobedience event in the capital city

The most effective act of mass civil disobedience is to have a significant number of people (at least 5,000-10,000 initially) occupy public spaces in a capital city from several days to several weeks. Again, this is not so much because of the material disruption that is caused in itself but rather the symbolic impact of mass sacrificial actions at the heart of national power, the site of the governing and media elites.

The precise design for the event will need to be made on the basis of the numbers willing to take part, the political context, and the expected police response. Groups should aim to get 20,000 people to central points near government buildings for several days.

There are two basic designs:

1. They would occupy public parks and bring tents to stay overnight. Each day there would be a public programme of discussions and assemblies, workshops, picnics, speakers, and entertainment. Think of a festival! On day two they would occupy several key road junctions for a day – at least 1,000 people at each one. The strategy depends on the police allowing this to happen (see analysis below). On the third day they would return to the junction and stay there permanently – putting up their tents, with a continuing festival programme.
2. A second option is that people would occupy the road junctions in a central location in the capital city from day one and stay there until everyone is arrested.

This then sets the scene for mass transgression; an act of mass sacrifice and a major public drama. The symbolic interpretation is people versus power on an epic scale.

Examples from history

The Leipzig 'Monday demonstrations'

This strategy draws upon seminal research[8] on the escalation of demonstrations at Leipzig in East Germany in 1989. The broad progression of events went as follows. A pastor was extremely disillusioned with the regime and led his community to hold a public demonstration over a weekend. Approximately 6,500 people attended. Initially, the local security forces did not interfere with what they considered a small Christian demonstration. Encouraged by this success and police reaction, the following Monday 17,000 people came out on the street.

Unsure what to do, the authorities contacted their superiors in Berlin. A message came down the line to shoot at the demonstrators. However, the next Monday there were 60,000 people on the streets and the police could not bring themselves to shoot that many people and thus disobeyed their orders. The following weekend 105,000 people turned out. The fear had gone, and this is the moment when the tide turned and, shortly afterwards, the regime. The lesson is that the regime was caught off guard by the non-linear increase in the numbers taking part in the demonstrations and the slow, centralised, top-down decision-making structure failed to act in time.

The design aims to catch the authorities off guard. They might allow occupations to continue for three days because it is not a critical issue. In this time the demonstration effect of thousands of people peacefully breaking the law to force government into serious action or stand aside, encourages thousands of people to join the direct action, or consider doing so. We aim therefore to hit the holy grail tipping point which leads to regime surrender.

If they allow mass civil disobedience to continue, then protesters will come to the capital and there will be unacceptable mass disruption. Alternatively, if they try to arrest, say 1,000 protesters in a day then the media interest will result in millions of people hearing about the arrests and the pleas to join the nonviolent mass action. Only a tiny percent (less than 1%) of these onlookers need to be mobilised for the net numbers (those joining minus those arrested) to rapidly increase. The more they arrest the more people join up.

Children's March 1963

Another inspiration is the American movement for black civil rights in the 1960s which organised a number of dramatically successful campaigns. One of the most famous was the Birmingham Alabama campaign of 1963 which shows key elements which are promoted in this booklet. The aim was to desegregate the city through a nonviolent direct action.

After failing to make an impression with Martin Luther King going to jail and the subsequent difficulties of mobilising the adult black residents of the city, the idea was hatched to involve the city's children and young people in an ongoing escalation of mass participation civil disobedience. The dilemma action design* involved thousands of pupils and students leaving school to illegally march through the centre of the city. Word was spread via local radio stations and meetings which promoted the methods of civil disobedience.

A 'D-Day' was set when the mass action would begin. The authorities opted for a repression response, arresting 1,000 protestors on the first day and 3,000 on the second day. This triggered a classic example of the backfiring effect. Thousands more children left their classes to protest and fill the jails.

'The fear had gone', as one officer reported. After a week there was no end to the mass protest. The authorities had lost control and the opposition collapsed. Decades of segregation policies were overturned in a week. This is the power of mass civil disobedience. The classic film The Children's March (viewable on You Tube) is required watching to see these explosive dynamics in action.

---

* A dilemma action is a type of nonviolent civil disobedience designed to create a 'response dilemma' or 'lose–lose' situation for public authorities by forcing them to either concede some public space to protesters or make themselves look absurd or heavy-handed by acting against the protest.

Jana Andolan II, Nepal

One of the most recent examples of successful nonviolent civil disobedience is the Jana Andolan II (People's Movement II) of Nepal. In 2005, King Gyanendra sacked the government and assumed direct powers, creating an absolute monarchy. He did this ostensibly to put an end to the ten years of civil war which saw 17,000 fatalities in a population of 24 million.[9]

The King's move was opposed by multiple social and political groups and led to the formation of a new and wide political coalition, including the largest active political group, the Maoist insurgents. With the formation of the new coalition the Maoists agreed to a temporary ceasefire.

The new coalition went on to mobilise an estimated 5.5 million Nepalese in civil disobedience and general strikes across Nepal lasting 19 days. The coalition included: civil society organisations such as the Citizens' Movement for Democracy and Peace, the Professional Alliance for Democracy and Peace and within these organisations, lawyers, teachers, engineers, professors, doctors, and journalists; NGOs; a coalition of four trade union confederations; the Maoists and a range of women's groups and peasants.

The focus of the protests was the paralysis of the cities, particularly the capital Kathmandu. On one day hundreds of thousands of protestors occupied the 27 km ring road and effectively encircled the city.

The protestors' demands were for:

- A return to democracy.
- Lasting peace.
- More inclusion of marginalised groups.

The protestors' tactics included getting high profile protestors arrested to encourage greater participation and creating spaces within the demonstrations for marginalised groups to speak and be represented. The government responded to the mass protests with a ban on mass meetings, assemblies and rallies, by cutting telephone and mobile connections and shoot-on-sight curfews.

After 19 days and 15,000 arrested, the King conceded and allowed the political parties back into politics. This ultimately led to not just new elections but also to an amended constitution effectively abolishing the monarchy. In 2007 a Comprehensive Peace Agreement was agreed and in May 2008 Nepal was declared a republic. Nepal is now considered to be more politically and economically inclusive of previously marginalised ethnic and caste groups. Since 2006 there has been no return to civil war.

Bringing the regime to the table

When the authorities lose the ability to stop mass mobilisation the regime is forced to negotiate. The point of the build-up design is that it minimises the chance that the police will shut down the civil disobedience before it has reached a critical mass of publicity. Moving a mass of people without the interference of the police requires a sophistication of coordinated management which will be difficult. However, if the numbers are available it will be better to get to the locations, sit down and stay there.

Alternatively, there could be a compromise between the two scenarios. Such considerations would need to be decided on the basis of the local context. The general aim however is clear: the closing down of the centre of a capital city day after day through the peaceful blockading of streets by thousands of rebels. Once this is achieved, we are in the ball park of forcing the regime to come to the table and talk.

Gathering the necessary numbers

A key factor is the numbers involved and this is why our whole strategy has to be centred around a primary objective – getting thousands of people into capital cities on a specific date. Success could be built by as little as five to ten thousand people. It is this group which potentially provokes the authorities into a repressive response which then brings many more thousands to join them on the streets out of feeling of solidarity and excitement.

The key is getting to the tipping point when coverage in the national print media and then the social media triggers this spontaneous additional mass mobilisation. This is how it takes off – when the fear has gone, as shown by our historical examples.

Make actions feel inclusive and fun
The inclusive and human feel of these occupations is essential to maintaining morale and nonviolent discipline. We know that the inclusion of children and older people can be extremely effective in stopping macho or aggressive behaviour whilst creating greater media interest. An occupation should not have the logos of other groups or parties but should be a bright display of Extinction symbols and whatever appropriate creativity the art groups produce.

Occupations can have a programme of events which could include the group sharing food and coming together in Affinity Groups for mutual support and human contact. At the same time there should be a programme of entertainment whilst being aware of the delicate balance necessary when promoting different cultural identities.

The general atmosphere should be 'we're going to take down the government and have fun doing it'. If and when aggression comes from authorities, it is responded to with humour and good grace. As the Children's March showed, this is possible even in a dangerous situation. Once a cultural and positive psychological frame is created in a large group it has a very strong socialising influence on individuals in the space and, vitally, on new people coming into it.

Publicity
Any large-scale event of mass civil disobedience needs to ensure that every opportunity is taken to promote the framing of nonviolent respectful action. The symbolic and sacrificial aspects of the law breaking need to be explained. This means giving out tens of thousands of leaflets in public places which explain the crisis and why breaking the law is now necessary, apologise for the disruption, and give information on how people should join. These could be handed out at bus stops and transport stations, parks or other public places.

Fly-posting should be done by several teams each night for the two weeks before the start date. 'There is going to be a rebellion against the government – join us', with specific details of where to go. The greater the diversity of posters the better to show the bottom-up nature of the mobilisation. Hundreds of people might stencil and paint the symbol all around the poor/inner city areas of the capital in the nights leading up to the event. People can spend the day on tube trains shouting out 'There is going to be a rebellion!' and handing out leaflets.

Super-noisy marches around sympathetic neighbourhoods of the city would effectively spread the message to people. This happened before the Tahrir Square occupation in Egypt during the Arab Spring. A peak in social media activity should be planned and an online campaign run to promote and reflect offline actions. An atmosphere of anticipation and excitement is created. After decades of defeat and cynicism millions of people begin to sense a real possibility for change.

We are in a new era where social media can be highly influential. Therefore, activists need to stay up to date and understand Facebook algorithms and how one bad Tweet without context can result in a negative news story. This kind of expertise is difficult to find and expensive to buy but activists must do their best either by reading the latest literature on social media marketing or asking experts to train team members for a day.

'Not Guilty' on all counts: Criminal damage at King's College
Social repression involving the climate disaster has shown up in many ways and this includes the court system. Two years ago, myself and a group of students painted messages on the walls of King's College London where I do my PhD research. Our aim was to persuade the College to divest from all fossil fuels, something they had refused despite four years of conventional campaigning. Petitions, meetings and sitting on committees had gone nowhere. When we painted the walls of the Great Hall of the College, we won in five weeks. Direct action worked.

I was immediately suspended and banned from entering the college. However, I openly challenged the ban to the point of getting carried out of the Students' Union by security staff and after ten days they removed the ban. The embarrassment was too great. Instead I was invited into negotiations with the vice principal and I then told him I was going on hunger strike until they had a signed statement committing the university to total divestment by a set date.

I was calm and respectful in all my dealings with him and he responded by initiating emergency investment committee meetings. Only by creating a crisis will institutions act upon an emergency. On the 14th day of my hunger strike a document was produced making the commitment to divest by 2022 and become carbon neutral by 2025. It was signed in front of the press and we all shook hands. We agree to take no more action against King's and the vice principal agreed no disciplinary or legal action would be taken against us. It was all very amicable.

A year and half later the crown prosecution service decided to prosecute me and another student for 'criminal damage without lawful excuse' – for painting the walls. We appeared in front of a jury and represented ourselves. We were told by the judge that the trial has nothing to do with climate change. He interrupted me 15 times and told me to stop talking about it (something I kept 'forgetting' to do).

For the judge it was a clear and simple case. We put paint on the wall, it is against the law and thus we are guilty. However, we argued what is obvious – our case was about climate change. It was about preventing the terrible suffering that will be created by the criminal fossil fuel industry unless there is wholesale divestment and it is closed down. We had a right of necessity to cause disruption in order to prevent massive disruption. This is a no brainer.

The judge couldn't get his head around it but the jury, ordinary Londoners, certainly did. They considered the case for the minimum time necessary. All of them then came to the unanimous verdict, we were not guilty on all charges of criminal damage. The judge told us 'you are free to go'.

This is yet another instance of the increasingly extreme social repression which we face in our societies at the present time. Everyone knows the situation is fucked – everyone except the political class and elites. The judiciary live in a make-believe world of formal legalities – they have 'lost the plot'. The advice from the pre-trial judge was that there was objectively 'no case to answer'. We were not guilty.

This total opposition of world views between the elites and the people is going to explode. Our job is to bring about this 'correction' in an ordered, nonviolent way through mass civil disobedience – but one way or another it is coming, racing down the tracks.

Building alliances

Join the Rebellion!
Potential political and cultural allies should be lobbied and briefed in the months before the Rebellion. The message should promote the three key motivators for revolution in contemporary western democracies: the accelerating climate and ecological crisis, extreme inequality and corrupt politicians and governments. This is a crucial movement-building challenge and is vital to mass mobilisation. The Rebellion has to morph at the last moment into a general rebellion against 'all government failures' in order to catch the regime off guard.

This prevents any opposition from framing us as just 'environmental protestors' and therefore being dismissed as a 'special interest' group. A key miscalculation which elites repeatedly make with uprisings is that they think it is just about one issue or group, not anticipating how rapidly it can change into a general rebellion against the all-encompassing illegitimacy of corrupt elite power. Our plan must include actions to proactively promote this overarching narrative.

Decentralised direct actions build the credibility with which to approach potential allies. It is not a matter of just presenting arguments but of showing you walk your talk. Movement-building actions should take place in all the cities and towns around the country, aiming to build a 'movement of movements'. Hundreds of local groups can then approach local potential allies with offers of talks and joint workshops.

The approach should be that we all need to work together because we all know the political system is not working and now it is going to take us to our deaths with the climate crisis. We need to rebel before it is too late. Once we have established the credibility of our strategy and ways of working, we can build support through two key developments: working with other groups and organising People's Assemblies.

Approaching NGOs and political groups

First, we can work with allied non-governmental organisations (NGOs) and radical political networks to hold joint actions and occupations. This would build the narrative of the connections between the ecological catastrophe, gross inequality and corrupted democracy. Practical cooperation and joint trust-building local actions will build the foundations for the strategic plan of the 'movement of movements' Rebellion in the capital city.

People's Assemblies

Second, and more structurally, we need to organise hundreds of People's Assemblies on the theme of 'How to solve the ecological crisis' or more simply 'What is going on with the climate crisis and everything else?' People's Assemblies are similar to Citizens' Assemblies, but without the process of sortition. Experts from around the world can help train facilitators and support groups toward issues of central importance.

The assemblies have an attractive format which draws in ordinary people. They are based around short two-minute personal testimonials and smaller break-out groups through which concerned citizens will create a new vision of the social and political transition needed to address the crisis.

Taking over large public spaces to have these assemblies would create dramatic political theatre, particularly if they happen in the context of disruptive and sacrificial direct actions. All this new bottom-up political energy would be funnelled into a new revolutionary political agenda, 'We have had enough of politics as usual, and we want a new political system'.

This new political awakening is legitimised through the demand for a National Citizens' Assembly, a formalised gathering of ordinary people selected in a systematic, randomised way from all sectors of the general population. The framing could be: 'The people have better ideas about what's needed than the politicians, so let the people decide'. A national assembly could then be set up to provide a living example of how we can organise our political life in a way which is truly democratic and inclusive.

Political allies

Environmental NGOs – such as Avaaz, Friends of the Earth and Greenpeace – are potential partners for mass civil disobedience actions. However, if they do not initially come on board, it may take a well-designed direct-action campaign which minimises the chances of alienation (by finding common ground) while maximising the chances of cooperation. This will be done by a 'maximum respect and maximum disruption' process.

The framing is positive: 'They have done great activism in the past, but now it's time to change', combined with short but noisy occupations which escalate, combined with media-attracting symbolic actions. These actions should be linked to a specific demand: that the institution or group declare a climate emergency and provide specific resources for mass civil disobedience. They can do an allied action at the same time and/or join in the main Rebellion activities. The time for joint meetings and joint statements is over. We need concrete commitments to support a Rebellion.

The key aim here is to create a domino effect. The Independent Workers Union of Great Britain (IWGB) achieved this in their confrontation with exploitative courier companies for better wages. They used nonviolent disruptive direct action to bring the first two companies to the table. After that the other companies agreed to pay increases without the need for direct action. Choosing the nearest potential allies first is the ideal starting point, then move onto the rest.

The starting point then would be the NGOs with their massive resources and promotion of methods of campaigning, which are now discredited as being ineffective and too slow. Once several of these are on board and they adopt the new strategy of promoting the Rebellion then we can move to the next ten or twenty organisations, doing quick disruptions of their events to show we are serious. The objective might not be to recruit the top management but to encourage radical change within their organisation from staff and members eager to join mass civil disobedience.

A similar process of creating allies, and/or creating disaffection through direct actions, should be organised for left-wing and other progressive organisations. Again, the message should be: 'The time for conventional politics is over, we are in an emergency, rebellion is now necessary.' A combination of disruptive action and an uncompromising demand is the key to success, not more meetings.

Local councils and other civic institutions

Other alliance opportunities exist with councils and civic institutions declaring a climate emergency. This should be seen as the beginning of the end of the old regime, not as a paper exercise as it is at the present time. The hypocrisy of accepting the truth but not taking appropriate action only makes these institutions more complicit in the criminality of the political class. The way forward here is to call on councils to give central government an ultimatum to call an emergency and act accordingly.

When nothing happens, they should be asked to break off administrative and financial cooperation with the genocidal regime. Council leaders should go on hunger strike to show their horror at the inaction of central government, people could withhold part of their tax or refuse volunteer roles that many local governments rely on. Such dramatic developments will attract the attention of the city and regional press, particularly when combined with occupations of council meetings and administrative offices.

The message is that if conventional political institutions do not respond then parallel institutions will form which will take action. Councils that then do come to understand that this 'emergency' is what it is – an emergency – can initiate dramatic measures in their localities to move to zero carbon economy.

These local political dramas will raise the movement's profile around the country and thus increase recruitment for a central mass civil disobedience event.

Media allies

Another critical network to reach is the media. Some outlets will be natural allies, and this might include a proportion of the radical and independent online press. However, the mainstream press should be the focus: the BBC and The Guardian in the UK, and the centrist and national media in any other country are likely to support, however the goal is to go beyond and reach all media outlets. We must go beyond The

Guardian! The aim here is to ask them to declare a climate emergency and then use that declaration to push for support for a rebellion explicitly. Suggest journalists and editors go on hunger strike on the main day of mass action. They should be asked to cross the line and go into existential conflict with the genocidal governing regime as a matter of national duty.

These extreme and unreasonable demands create a crisis in these spaces and widen the Overton window,* as it has with the use of the words 'extinction' and 'rebellion' over the past year. It will become clear that their role in this emergency can no longer be just writing about it. The sea change in approach can be aided by asking for private and informal meetings with journalists, editors and media people to give them the standard talk which will help develop contacts with those who will support the Rebellion.

This approach should then extend to the right-wing press – using the framing of 'order, security and legacy' (see more on pp.55–61) as a way to highlight the contradiction of believing in these values and yet having no concern for the emergency which presents an existential threat to them.

Cultural allies

Any cultural group or institution in society can be approached – the climate crisis affects everyone and everything. People can stand up in cinemas, theatres, conferences, lectures, trains, supermarkets, restaurants etc. and declare that we will soon die if nothing is done and call on people to join the Rebellion. We need to approach religious and spiritual organisations, small businesses and community groups. The possibilities are enormous, and this is the agenda for the mass movement we are creating.

The actions can be videoed and live-streamed and combined with trainings so people can select their own priorities without any central direction, other than to follow some basic guidelines. Local groups can brainstorm and select groups they want to approach.**

---

* The Overton window is a term for the range of ideas tolerated in public discourse, also known as the window of discourse.

Schools, colleges and universities

A particular focus should be schools, colleges and universities. The mobilisation of young people is already happening around the world, but needs to move to more prolonged and disruptive forms of civil disobedience. A possible process is as follows: student campaigners hand out leaflets outside their place of education promoting a strike and civil disobedience event. They enter into conversation with people as they hand out the leaflets: 'Could I speak to you for a minute about the climate crisis?'

A script can be used to guide the conversation into listening to what the young person thinks about their future. Contact information can be taken from the most enthusiastic people who are incentivised to organise further outreach and engagement.

This would lead to the recruitment of more recruiters to canvass and arrange meetings with the school, college or university. Through this process the young people start to organise their own direct action and mobilisation, with training and mentoring from experienced rebels. Youth events would happen separately from those for adults in the run-up period to a central action. Ideally, they would be integrated within the wider coalition for the main mass civil disobedience event.

There is a great opportunity to create real diversity in Rebellion mobilisations – children and young people from inner city schools mixing with middle class campaigners from the suburbs. Serious resources then should be devoted to creating city youth mobilisation.

Local community outreach to extend diversity

A similar diversity priority would be to develop local community meetings in inner cities. The process would involve leafleting and then door knocking. This has been tried but still needs to be developed and standardised. As with youth mobilisation, the 'ordinary' people who are mobilised through this process should set up their own groups and networks so that they are not put off by the specific culture of middle-

---

** The self-organising system and non-hierarchical, non-centrist aspect of XR is effective and powerful, see 'Structure of XR UK' on YouTube: https://www.youtube.com/watch?v=qZFR7Uia4-o

class or radical campaigning, but are integrated once they are established into a central, major mobilisation.

All this should be combined with a nationwide campaign to roll out the standardised talk, 'Climate breakdown is happening, we are heading for extinction and this is what we can do about it'. Again, the focus should be on local communities rather than just green political groups. A target should be set for how many people we want to reach, and various mandated working groups should create plans for how to achieve this (e.g. on doing trainings, canvassing etc.).

Lastly, national telephone campaign teams should be set up to reach people on the database to tell them about the details of the mass civil disobedience. One-to-one approaches are critical to persuading people to take part in the Rebellion.

Framing and messaging

Mass movements are successful when people who hate each other join together for the common good.

Chris Hedges (paraphrased), ex-New York Times journalist, who covered revolutions and uprisings for 20 years.

This booklet is primarily focused upon the organisational and action design for societal and political transformation. The other side of the equation is the way in which we communicate what our demands are and why it is vital we succeed. Messaging and framing are central to any plan for success.

Reformist framing of solutions must end

The bit-by-bit reformist framing of change is both immoral and ineffective as it puts political ideology before scientific facts. This may be justifiable if we are in a period of reformist possibility, but that window is now closed. We are now in a new political context where telling the truth is both effective and moral. This is a switch from a degenerative post-modernism where 'presentation' takes priority over actuality. It is replaced by a new realism which now provides the best path to create the massive structural changes that are required. The key example here was the decision by Extinction Rebellion to make 'Tell the Truth' their first demand. This kick-started a social movement.

In terms of the actual threat to life, this new approach enables us to engage in a straightforward risk analysis which happens routinely in other areas of social and economic life but not in the area of climate change. This means taking a scientifically measured level of the threat and then multiply it by the probability that it will happen, which gives a risk assessment.

Of course, the method cannot be totally accurate, but we have no better means of assessing risk. In the case of climate breakdown we are looking at a high probability of the death of billions of people in the next generation or two. The reformist denial of this reality is catastrophically irresponsible. In the last century such levels of denial in facing the truth have led to appalling atrocities of communism and fascism.

Telling the truth as a revolutionary act

This simple act of telling the truth allows the Rebellion to engage the media with the narrative that they are simply acting appropriately in the face of the same ignorance and denial that led to World War Two. Denying the reality, we face 'climate appeasement'. This means insisting on speaking the truth without interruption during interviews and insisting on giving time for listeners to process the emotional impact of realising the extent of the terrible political corruption that claims to be 'solving the climate crisis'. A key approach to communicating this reality is to use direct actions when speaking to the media.

Some options are:

- To walk out of the interview after making a short statement.
- To insist on silence after telling the interviewer we are heading towards mass extinction and that we are all going to die unless urgent action is taken.
- To simply repeat a prepared short sentence about the reality we face over and over again regardless of what the interviewer says.
- To refuse to leave the studio until removed by security as a protest at the media coverage of climate change, but to go peacefully when removed.

All of these moves aim to prevent journalists and presenters avoiding the true horror of mass extinction. Actions speak louder than a thousand words, and we need to see the media as a place for system change rather than for pseudo-rationalist debate about false gradualist improvements. Without painful emotional engagement there will be no real commitment to drastic action, and thus no real change. Only when the rules of media engagement are broken can we create effective communication that reaches hearts and minds.

That said, if and when we are given an hour to carefully outline the full magnitude of mass extinction, and the social scientific evidence for Rebellion as the most effective response, then we should of course take this opportunity if a journalist or editor is willing. We are not against having a serious discussion, but rather against spaces which have no interest in such discussion.

Media and messaging working groups should have the remit to recruit speakers and train them in disruptive action – which sends out what should be our central message: 'We are different from them (the status quo or regime), we are not afraid of them and you don't need to be afraid either. Rise up and exercise your freedom. Come down to the occupations and become a human being, not someone living within a media and politically constructed "post-truth" world'.

These media appearances have the potential to go viral and thus can have as much influence as the civil disobedience actions on the ground. This new opening for direct action is then a vital additional front which needs to be central to the overall strategy.

Internal messaging – the value of diversity

Research shows that members of a social group or network tend to have irrational 'in and out group' emotions and reactions that consciously or unconsciously prevent them working with people who are not like themselves. This problem is unlikely to resolve itself easily. To build a mass movement we inevitably need to bring together people who ordinarily don't want to mix. There are several effective approaches to this central problem of effective mobilisation:

1. Make all campaign spaces as friendly as possible on a personal level. This means maximising face-to-face human connection as a counterbalance to any perceived alienating group identity. People should be explicitly welcomed into the meeting space. They should then sit in small mixed groups in meetings and be given time to get to know each other. There should be food and drink at every event to ease social interaction. There should be calls or one-to-one meet-ups afterwards to thank individuals for coming and to listen to any concerns. People should be appreciated often and generously. Any criticisms should be restricted to critical problems. All this is possible and an essential part of overcoming the conflicting identities problem.

2. The main campaign spaces need to be as 'culturally neutral' as possible. This mainly means taking out subliminal social class or mono-cultural elements. This can, ironically, mean changing processes which are supposed to create inclusivity but actually exclude people. These include someone telling the group everyone is welcome (people feel welcome through one-to-one welcoming behaviour, not through being told something is true because someone says it is).

   Activist routines can be off-putting; for instance, the hand signal conventions of Nonviolent Communication and new age or academic (radical left) language. Everything written and spoken should be put into commonly known phrases e.g. 'the way someone talks' rather than 'discourse'. People being overly physical and friendly with each other whilst ignoring newcomers (creating the impression this is someone else's group that I am entering) should be discouraged.

   Sources of division such as social class, race and gender present very real challenges. It is why working-class people are almost totally absent from UK environmental movements. Poorly constructed messaging will put off new people from another identity not least because of the prejudices of the people receiving the message. Some of this may be unavoidable. For instance, a black woman came to an Extinction Rebellion meeting and left afterwards intending not to come back because there 'were too many piercings'.

3. The structural solution is then to create different spaces for various different groups. For example, working-class mobilisations are organised by working people themselves (as opposed to middle-class groups that claim to speak for them). Similarly, people of colour can organise in their own spaces. People should therefore be encouraged to set up their own groups which agree on basic red lines such as nonviolence but are able to promote their own cultural identities.

These internal organisational issues are critical to building a mass movement and so move the environmental movement out of the middle-class bubble that has defined it for decades. As the research[10] shows we need to create mass participation civil disobedience – this means we need to engage with, and mobilise, many diverse cultures. It's not about creating a comfort zone but about getting on with the critical work that needs to be done – it's not going to be easy, but it has to be done if we are serious about succeeding.

External messaging – the value of inclusivity and universalism
This is one of the most difficult areas for green and left-wing activists to understand and accept, but it is critical to political success. We must appeal to people who don't join or support environmental causes, be that because of ideology, social class, culture, religion or race. This has been done with massive success by the political right wing for decades in order to demoralise and confuse left-wing supporters.

They take a left-wing idea, or word, and co-opt it to right-wing purposes. 'Revolution' is an example – or the idea of workers sitting on company boards which Theresa May promoted on the day of her national election victory. Of course, there was no chance that was ever going to happen.

Left-wing movements can do the same but ideological purism often prevents them from being as creative. However, with the climate crisis and ecological breakdown there is a unique opportunity to play the right at its own game because in this case the arguments are genuinely universalist. There is then a massive opportunity to build up right-wing support and/or demoralise the opposition by parking our tanks on their lawn (to use a right-wing phrase).

The framing should be to ditch environmentalist language and adopt the language of traditional liberal universalism. This was done to great effect with Extinction Rebellion's Declaration of Rebellion and the letter to the Queen, delivered to Buckingham Palace in November 2018. In no sense does this explicitly exclude a left-wing orientation. Notions of honour and duty were widely respected values on both the left and the right until neoliberalism reduced everything to self-interest and monetary value.

Notions of left-wing national identity and civil nationalism have been central to many traditional mass left-wing movements. Words like honour, duty, tradition, nation, and legacy should be used at every opportunity. Not only is this language new and therefore attracts attention but it can be connected to a profoundly egalitarian ideal. In fact, historical research[11] has shown that inequality is usually reduced not by left-wing administrations but by governments facing national crises such as war. In these circumstances the taxing of the rich is seen as a universally accepted necessity, as it should be by any regime realistically concerned with addressing the climate emergency.

Progressive solutions are desperately needed

The tension here is that for the past 30 years left-wing and environmental movements have had neither the structural opportunity nor the creative innovation to radically challenge the fundamentalist neoliberal regime, where the real power and capability to exploit natural resources exist.

For instance, 'climate justice' movements have been keen on declaring solidarity and morally 'good' attitudes but have had no practical or credible plan of action other than calling out the government's lack of action whilst going on marches and signing petitions. If they were serious, prisons would be full of people following through on their outrage at the terminal destruction of the natural world for profit. Instead, when a movement comes along which undertakes mass direct actions which actually do directly challenge structural inequality, they seem more concerned with the words and statements of the movement rather than real world successes.

As such these 'activists' ironically do the work of the neoliberal elite they claim to be vigorously opposed to, by undermining any effective action by engaging in endless judgement and criticism. In the meantime, many people follow the NGO approach of 'armchair activism' or 'clicktivism' whereby individuals think they are making a difference by signing an online petition when in reality their concern has little to no impact via this medium. However, this form of activism can lead to 'on the ground' activism. People who have signed petitions should be contacted and invited to join the Rebellion.

Radical direct action is by definition an exclusive act – not everyone can or is willing to do it. But it is also the only way that structural change happens. We need to make this argument to our 'radical left' critics and not allow ourselves to be pulled into the ghetto of excluding 'inclusivity'. We should be speaking a new universalist language, using Martin Luther King's speeches as a prime example of how to reclaim the framings of national pride to build a broad mass civil disobedience coalition.

The points here are difficult but there is no question where effectiveness lies. The task is to reiterate again and again that we need mass, high-participation civil disobedience concentrated upon a single event – a Rebellion. If we are serious about the truth we face with the ecological emergency, we have to be equally serious about organising and rebelling effectively. To do it well and thoughtfully is our moral duty.

Post-Rebellion done right

Citizens' Assemblies

The key to success post the Rebellion is to channel an explosion of radical political energy into a legitimate means to make collective national decisions. The brilliance of the mechanism of Citizens' Assemblies is that it appeals to both liberal and revolutionary values. It is do-able as is demonstrated by its use over the past decade or two.

It is deeply democratic and popular (involves ordinary people) so that no democrat or liberal can object to it – as opposed to the danger that a revolutionary left-wing elite removes democratic processes in a bid to create a socialist society (been there and done that!). At the same time, it keeps the revolutionaries on board because it creates a forum where informed deliberation and reason will finally be given space to trump the power and corruption of big money.

And for good reason, we can predict that the outcomes of assemblies will be far more progressive and rational than conventional, cynical commentators would predict; 'No, we don't want the rich and powerful robbing us of the fruits of our labour, and no, we don't want our children to die in a climate catastrophe ... thank you very much.'

The proposal then gets the best of both worlds. It proposes a clear and credible solution that is supported by the social scientific evidence.[12] We have established that when a society reaches a point of extreme imbalance then only a revolutionary episode can be successful in reorientating it. This reorientation is concretised through a replacement of the usual first-past-the-post democracy with a sortition system that comes to shared decisions, and as such provides a clear pathway to a post-revolutionary resolution to the central question of politics: 'Who decides?'

A solution which is satisfactory to both liberal and revolutionary constituencies is necessary. It is only when these two political orientations make an alliance that an elite falls. A key part of our strategy will be to sell the plan to a critical mass of the liberal elite 'defectors' as well as to the 1% of the general population who will lead the disruption.

Citizens' Assemblies in more detail

Citizens' Assemblies, chosen by the process of sortition (i.e. randomly selecting citizens from all sectors of society), have become very popular over the past few years, especially regarding how Brexit could have been deliberated over more effectively. In essence, they are the roots of original notions of democracy. As part of a Citizens' Assembly, the selected citizens are exposed to a 360 degree understanding of an issue, which puts them in a better position to make decisions than politicians who are under the influence of a barrage of lobbyists and careerist considerations, both of which take them away from the simple matter of making an educated choice that is aimed at the best outcome for all.

Citizens' Assemblies allow ordinary citizens to learn about and then decide issues that are often too hot to handle for politicians who fear repercussions from the electorate. The Citizens' Assembly in Ireland* in 2016 is a good example where an issue that could have destroyed a political career was taken on and deliberated over by 99 randomly chosen citizens to great effect and succeeded in repealing an archaic law that demonised women and took away their right to make decisions about their own bodies.

In the case of climate breakdown and how society is going to avoid the worst effects of it, Citizens' Assemblies are our only truly democratic hope. The transition that will be shown to be necessary would be political death for any one party should they suggest the changes that will be required. It is the antidote to the corporate-captured broken democracy of today and indicates a future that can be run truly by the people for the people with the well-being of all as the central operating principle.

---

* For the Citizens' Assembly in Ireland see https://www.electoral-reform.org.uk/the-irish-abortion-referendum-how-a-citizens-assembly-helped-to-break-years-of-political-deadlock/

A transition of power

The post-revolutionary plan then needs some detailed working out to be credible. I suggest there could be a transition period leading to a permanent new political constitutional settlement. For this a professionally created and transparent National Citizens' Assembly (NCA) would be established containing, maybe, 1,000 people for a fixed period of two years. It would then create regional and city Citizens' Assemblies to facilitate the decentralisation of power.

The National Citizens' Assembly would deal with social and political legislation, enacting emergency measures on the climate crisis. It would also create a new written constitution which would ensure such assemblies were a permanent fixture of our political life.

Continuity is king

The British political tradition of avoiding mass bloodshed at such moments of structural change could be continued by creating a semblance of continuity in the following manner. In the transition period Parliament could still exist but with a proportional representation system. It would then have an advisory role to the NCA in a similar relationship to that which the Lords has with the Commons at present.

A process could be set up whereby the NCA would produce a piece of legislation, with help from policy experts and lawyers, which would then be sent to the Commons for debate and approval. If it was rejected, then the NCA could send it back again after three months. This process could then be repeated. Then the NCA would have the final power of enactment. The critical design point here is that the final power resides with the NCA.

The idea is to have some 'dignified' part of the constitution as developed in the 18th and 19th centuries. In the 18th century the monarch was still technically in control, but real power passed to the aristocracy. Then the Commons took control over the Lords and became progressively more democratic during the 19th century. We should provide a way of satisfying the not-to-be-underestimated desire for ritualistic and traditional continuity, enabling real power to become democratised.

We can see a contemporary example of this process replicated through the creation of the most successful participatory political system in the world at Porto Alegre in Brazil. Neighbourhood assemblies elect one-year representatives to join together to create a city budget. A provisional budget is drawn up and then goes back to the assembly for feedback. Then a final budget is drawn up and presented to the conventionally constituted city council which always has to agree to the budget. Technically the assemblies have no constitutional power and the council is sovereign, but the councillors know that real political power resides in the people power of the assemblies and so would never deny their will.

There could then be a ceremonial role for the Lords and the Queen. It could be up to the NCA to decide on this. Keeping the Commons too could be necessary in order to overcome traditionalist opposition (not least from the Labour Party).

Pushing the Overton window: failure lays the foundations for success

On deciding on this strategy, there are two outcomes which are both very attractive. First is the classic revolution as outlined above. The government's power to address climate change is removed and the new people's power takes its place. However, there is a second scenario, which in the present context is still clearly preferable to any reformist orientation and that is what I would call the 'glorious failure'; the Rebellion does not succeed but it has a massively empowering effect on the national and international political imagination by shifting the agenda towards what can concretely be asked for and planned for – namely system change.

This is called transforming the Overton window, the range of ideas that are openly discussed in public debate. If failure enables previously unspoken possibilities and realities to be discussed, then we have not failed. In fact, we have made a critical contribution to the transformational changes that now have to happen.

Although we may 'fail', the baton will be passed to other runners in the team who will learn from our mistakes and win the prize (e.g. the French). We will have created a demonstration effect for an immensely attractive 21st century model of revolution – nonviolent, participatory, and genuinely democratic – compared with the dire record of revolutions in the 20th century. We need a way to transform societies with minimal violence and which maintains a balance between liberty and equality, and builds cohesive communities.

Learning from revolutions past

We know that a dogmatic pursuit of discredited revolutionary models can be socially ruinous. The Citizens' Assembly system answers the age-old question of 'Who decides?' and represents as big a political shift as the transition from aristocratic rule to representational democracy. I don't think the constructive effect of such a reframing of the revolutionary project should be underestimated.

Without it, we are left with directionless and spontaneous uprisings such as the one we have just seen in France with the nicknamed 'gilets jaunes' riots, which research shows usually lead to authoritarian outcomes and civil war.[13] It is easy to destroy a system but much more difficult to create a better one. The model outlined in this booklet gives viable solutions to both sides of the equation.

The context of the threat of imminent mass death and destruction is the final and decisive argument for this strategy. There is simply no chance of getting the rapid changes which are needed through negotiations which leave the present political class in power. This is both a terrible thing and a good thing. Terrible because if we do not succeed the consequences will be bad beyond our imaginations.

Good, because in the face of such a challenge, there are no longer any grounds for hesitation. Our purpose is clear. There is no alternative. In a paradoxical way then the clarity and purpose will make us happier than before. It is clear why we are here in this world and what we need to do with our lives.

The Great Transition

As Greta Thunberg put it: 'We must admit that we do not have this situation under control. We must admit that we are losing this battle. We must stop playing with words and numbers, because we no longer have time for that.' And David Attenborough said: 'No action is too radical.' These words echoed around the world; we must pay attention and act accordingly.

Arguments about the need to 'protect the economy', ignoring that the current economy is taking us over a cliff into collapse, simply cannot hold up. So, let's be clear. A Climate Emergency is not a rhetorical call for accelerated climate action, it's a call for a major transition of the economy. There is simply no other way to get the job done.

But what does a real Climate Emergency action plan look like? The answer is defined by the science – by the physics and chemistry. This is the focus of this chapter. Some find it hard to imagine this is actually happening. Some think it seems 'unrealistic'. But ask yourself about the alternative of risking the collapse of civilisation, the death of billions of people and the extinction of much of life on earth? Is that 'realistic'? No economy on a dead planet.

Delivering the Climate Emergency response

What do we need to achieve? Listed here are some of the processes and measures needed to deliver a credible Climate Emergency response.

We have drawn this from a series of expert analyses, studies and publications* over the past decade, including 'The One Degree War Plan' by Prof Jorgen Randers and Paul Gilding from 2010, which was

---

* For list of publications see p. 75

perhaps the first analysis proposing a WWII style mobilisation to return the earth to 1°C of warming. We also draw, amongst others, on 'The Victory Plan' from 'The Climate Mobilization' and a paper by Dave Roberts 'What genuine, no bullshit ambition on climate change would look like'.

It's not intended as a precise plan. It is to show what needs to happen and to establish that it is clearly a path we can choose.

Here's what we need to achieve:

Stop the world warming

We need to stabilise the climate at between 1°C and 1.5°C temperature increase above pre-industrial levels. We need to stabilise CO_2 levels at about 350ppm. That's the task. No further negotiations needed.

Doing so will require us to largely eliminate human created greenhouse gas emissions of all types within a decade or two and also take actions to cool the earth. The latter is required to reduce the risk of triggering runaway climate feedbacks or tipping points.

Eliminate fossil fuel use and close that industry down

We need to eliminate fossil fuels from the economy, and we need to do so within 20 years, with most of the work done in the next 10. That means immediately banning all new investment in fossil fuel exploration and development.

- Close down all coal-fired power stations – the dirtiest within 5 years, and the remainder within 10.
- Close down all gas-fired power stations, most in the next 10 years.
- Convert all transport to electricity, with the electricity generated by zero carbon energy sources.
- Manage this process with a massive reduction in energy use even if that means rationing.

This all means we will reduce the income of fossil fuel companies worth trillions of dollars, including all of the world's oil, coal and gas companies. They had the chance over 30 years to transform and chose not to. Now they must work with a transition process and reinvest in renewables or go out of business. Governments will have to provide education and retraining programs to help people who lose their jobs and address the impact on the communities. The financial implications for the national and local authorities and the pension funds who have investments in fossil fuels will also need to be addressed.

If global fossil fuel combustion is rapidly eliminated, the earth will experience a surge of warming due to a reduction in the polluting aerosols (which in turn reduce the level of sunlight reaching the earth). To counter this, we will need parallel drastic cuts in short-lived climate pollutants such as methane, black carbon, hydrofluorocarbons, and ground-level ozone.

Drive massive energy efficiency including rationing and demand management

To close down power plants, even with rapid global expansion of renewables, will need a massive global drive for energy efficiency and probably, energy rationing until we get there. It will, in most cases, be economically beneficial to drive such efficiency.

Restore forests and ecosystems

We need a massive global reforestation program, planting trillions of trees to absorb CO_2 as proposed recently in the journal *Science*.[14] This is one of the cheapest ways to absorb carbon from the atmosphere. It will take decades to start to absorb large quantities of CO_2, but reforestation will then have a huge impact over the following decades and will help to restore the climate, refreeze the earth's poles and be enormously beneficial to biodiversity.

Reduce the greenhouse gases in the atmosphere

We face tipping points that could trigger runaway climate change with the system spiralling out of our control and the likelihood of global collapse within a decade or two. We need to cool the planet as fast as possible. This means shifting from just focusing on 'reducing emissions' to also 'reducing warming' to give us the time for a parallel rapid reduction in CO_2 emissions to have its long-term impact. There are a number of ways to slow and then reverse short-term warming – including geo-engineering solutions.

But the cheapest, fastest and best understood action would be radical reductions in methane emissions. That means:

- We need to eliminate the use of gas as an energy source in less than 10 years.
- We need to dramatically lower consumption of animal products. Vegetarian and vegan diets are ideal, but everyone can simply reduce or eliminate meat and dairy consumption. This will come with a huge cost saving on healthcare.

Mobilise action from many sectors

Many more things need to happen, as described in the papers referenced at the end of this chapter. Most are well proven and just need to be taken to scale. People will argue the details, but if you want to succeed, it is necessary to accept a level of risk and have a level of confidence as to what a united humanity can achieve.

Some things are very clear and should be confronted now. There can be no more burning of fossil fuels in the world. Whether this should be achieved within 5, 10 or 20 years is for later. Whatever path we take means there will be no oil, coal or gas industry. Either they go or we go (and then they go anyway – no industry on a dead planet).

Measures to achieve the above

These are the outcomes we need. But to get there we will need a series of measures including mobilising the community, policy to ban certain activities, the application of taxes, subsidies and mandates by government. Some examples include:

- Carbon taxes and dividends both to drive behaviour but also to compensate the poor.
- Just transition strategies to manage the social consequences of rapid change.
- Taxes on all virgin materials to encourage investment in recycling.
- Feed-in tariffs to drive distributed energy and storage in homes, schools and factories.
- Announce a date for a total ban on fossil-fuel-powered transport of all types; land, air and sea.

- Suspend expansion of all airports and roads to send a clear signal as to society's priorities.
- Electrify everything we can and ration the supply of fossil fuels including petrol for cars.
- Mandate a compound 10% per year reduction in plane travel for the first 5 years and achieve this through taxes which can then be used to invest in alternatives including rail infrastructure.
- Incentivise soil carbon sequestration on a large scale by establishing a system to pay farmers and landowners for increasing soil carbon, which also enhances soil quality.
- Drive changes in diet that reduce emissions and enhance health through policy including pricing and public education like successful WWII campaigns and Meat Free Monday.
- Scale up standards for the built environment with zero-carbon housing and transit-oriented development. Subsidise energy efficiency retrofits in low income housing.
- Use demand management to match energy use with availability.

The economics: can we afford all this?

That's such an odd question. We face the collapse of the global economy and civilisation with the potential for the extinction of much of life on earth and we ask if we can afford to stop that? Do we think carrying on with collapse might be cheaper?

Yes, we can afford this, and it's actually not that expensive. Government has considerable financing powers to make sufficient funds available. As argued in Randers and Gilding's 'One Degree War Plan', a carbon tax at $20 per tonne, rising by $20 each year for 5 years to $100 would raise around $800 billion in the first year and around $4 trillion by year 5. This would be approximately 1–3% of GDP. That is just one of the many mechanisms available. Money raised should be used to fund the emergency response and to alleviate the resulting hardship – primarily among the poor.

Such an economic mobilisation has many potential benefits, including increased equality, higher employment and the integration and social cohesion of society.

A new transition movement: if governments don't listen then the people will act

Building on the existing Transition Network,* a major people's movement is needed to organise the mitigation of climate change, ecological restoration as well as preparation and adaptation to the coming changes. We need programmes like TransitionLab.earth that are focused on volunteerism at high levels such as engineering; actioning solutions rather than more policy 'think tanks'. For instance, inventors could research and develop battery technology and give away the patents for free. In fact all beneficial research needs to be part of the Commons, i.e. for free and unrestricted use.

An expert group of data scientists could develop artificial intelligence to help predict the worst affected areas of the UK, then engineers could design systems to mitigate the impact. Efforts like this are underway all over the world but there need to be central platforms that citizens can engage with.

A Lab should be created, made up of scientists, engineers and other Earth-enhancing experts to advise and respond to questions, and take direction, from Citizens' Assemblies. Based on this collaborative work the Assemblies would then go on to organise people into a Citizen Voluntary Service. Critically, the Lab would not be government funded or be a corporate entity. It would be a service for the people, by the people, rather than waiting for government to pass policy and commission agencies to work. The work starts now.

Rapid-transition working groups

Groups organised specifically to push urgent transition could focus on some of the following issues:

1. *The economy of change.* Economists, in consultation with renewable energy specialists, transport experts, biologists, natural scientists etc. will work on the cost of converting the world economy to avoid collapse. Where does that money come from? We could start with tax evaders, the reduction of military spending and fossil fuel subsidies.

---

* Transition Network is a movement of communities coming together to reimagine and rebuild our world, see https://transitionnetwork.org

2. *Preparing communities – the physicalities.* Flood-prediction projects are an example of communities getting prepared; they are starting in India as awareness grows about the Himalayan glaciers melting and monsoons becoming chaotic and extreme. But there will be other adaptation techniques that apply more relevantly in different parts of the world. Independent scientists and experts in the Lab would be part of a Wikipedia-type model of open, public ownership of the Great Transition.
3. *Preparing communities for resilience in the Great Transition.* The social implications of climate breakdown are immense. Work needs to be put in now to developing and maintaining sustainable communities, no one will survive alone. Our way of life will change not only from lack of the usual resources but also from the inevitable growing climate refugee crisis.
4. *Food.* How do we feed the world in the event of mass collapse of insects, ecosystems etc.? Red meat consumption will need to decrease rapidly and great effort put into improving our ravaged soils. The whole population will need to become more familiar with growing their food wherever they can, in addition to new farming methods and eco-enhancing systems.

None of these issues are being addressed by any trustworthy entity. Trust in the UN has reduced significantly globally, tech companies aren't doing enough, and governments are reluctant communicators. However, we have models: Vox, Now This and The Years Project are great examples of the use of social media to spread critical information about important issues. In the mainstream media, David Attenborough and Brian Cox have captured the imagination of the world on the subjects of science and protecting nature. The transition movement needs to learn from these communicators.

Global voluntary service for transition

In the worst-case scenario, we may need to start building on the great work of the Transition Towns initiative on a massive scale. This could be part of the work of Citizens' Assemblies which would organise a voluntary scheme for anyone who wishes to lend their skills towards some of the following:

- Flood defences
- Pre- and post-climate change planning
- Sustainable communities
- Renewable energy
- Disaster preparation, prevention and recovery

This is essentially the work of local councils and authorities. However, where gaps exist, a transition voluntary scheme would supplement and help.

The transition has to be imagined, visualised, planned and actualised. There is no time to lose. It will require new institutions – a parallel social revolution to the political revolution we need to transform the state. Again following 'business-as-usual' is no longer an option. We have to be realistic – everything has to change.

Conclusion

It is clear from the above, necessarily sketchy, description that defining what we need to do is relatively easy. The big issue is, will we choose to do it?

We must put our bodies on the line and refuse to accept extinction. We are not naïve as to what this will take. It will mean shutting down cities, shutting down companies and if necessary, shutting down parts of the economy. We will be forced to do this because the physics requires it. The path we are on goes off a cliff. Transition is the only way.

It is completely realistic to prevent extinction.

We plan to do so.

List of publications

The scenarios and actions outlined in this chapter have been derived from:

Ecofys/Navigant (2018). Energy transition within 1.5°C – A disruptive approach to 100% decarbonisation of the global energy system by 2050. White Paper. https://www.navigant.com/-/media/www/site/downloads/energy/2018/navigant2018energytransitionwithin15c.pdf

Hawken, P. (Ed). (2018). *Drawdown: The most comprehensive plan ever proposed to reverse global warming*. Penguin. See also www.drawdown.org. Summary of solutions by overall rank see https://www.drawdown.org/solutions-summary-by-rank. References, citations and resources see https://www.drawdown.org/references

IRENA (2018). Global Energy Transformation: A roadmap to 2050. International Renewable Energy Agency, Abu Dhabi. https://www.irena.org/publications/2019/Apr/Global-energy-transformation-A-roadmap-to-2050-2019Edition.

Jacobson, M.Z. & Delucchi, M.A. (2009). A plan to power 100 percent of the planet with renewables. Scientific American, November. https://www.scientificamerican.com/article/a-path-to-sustainable-energy-by-2030/

Randers, J. & Gilding, P. (2010). 'The one-degree war plan'. *Journal of Global Responsibility, 1*(1) 170–188. https://paulgilding.files.wordpress.com/2015/01/one-degree-war-plan-emerald-version.pdf

Roberts, D. (2018). What genuine, no bullshit ambition on climate change would look like: How to hit the most stringent targets, with no loopholes. www.vox.com. Updated 8 Oct 2018. https://www.vox.com/energy-and-environment/2018/5/7/17306008/climate-change-global-warming-scenarios-ambition

Silk, E. (2016/2019). The Climate Mobilization Victory Plan. The Climate Mobilization. (Revised March 2019 by Bamberger, K.) https://drive.google.com/file/d/0Bze7GXvI3ywrSGxYWDVXM3hVUm8/view

The cost of freedom is civic duty

I recently saw the film 'First Reformed', about a priest who considers blowing himself up in the face of the existential crisis of climate breakdown. The director, in an interview, casually asserts that the human race is certain to be extinct by the end of the century, before answering the next question on some mundane aspect of film production.

The film and interview illustrate the depth of moral depravity to which our society has now fallen. We collectively know we are about to destroy our children's lives but can only muster individualistic responses and casual nihilistic indifference.

Let's be clear: such moral degeneration is what enabled the Nazi death camps to happen – a complete collapse of any sense of empathy or duty to uphold universal human values and the most basic moral principle, 'Do unto others as you would have them do unto you'. Not, be nice to others just because we want good things to happen to us, but because it's the right thing to do.

This booklet proposes another way – a light between the darkness of apathy and cynicism. We have been here before – in fact numerous times. Many societies and cultures have experienced the prospect of annihilation and it is clear which values and actions need to be adopted if our chances of survival are to be maximised.

The first step is a return to some sort of transcultural balance between the individual, the society and the state. We have a duty to protect our society as much as to attend to our self-interest. We are all unique individuals but also intricately connected social beings. We will therefore stand together and survive, or die separately.

In the West, the discovery of this middle ground gave birth to our most glorious political discovery – the assertion of a set of universal rights and obligations which act as a bulwark against both individualist narcissism and murderous autocracy. In this booklet I assert, along with Thomas Paine* who inspired the title, that the people come first against any special interest – whether that is a tyrannous monarchy or a corrupted merchant class. The people have the right to decide, and in return, the people need to step up to their obligations. Rights and responsibilities go hand in hand.

And so again today there is a need for us to step up and make the sacrifices that our forebears made many times before us to win the freedoms we enjoy. We are about to lose them if we do not wake up and act. This will necessitate giving up jobs and taking time away from our families. It requires pain and suffering because no common good has ever been created without it – especially now as we enter our darkest hour. This is what growing up means: to see a situation as it is rather than how you would like it to be and to respond in a responsible manner.

A Rebellion, as outlined in this booklet, is not a consumer choice – not an act of possession or enactment of an identity. If it is, then it will fail miserably. It can only work as an act of universal service and duty – a fulfilment of our civic and republican tradition which pulled us out of the dark ages and lives of impoverishment. It is a rugged, anti-utopian liberalism that asserts the truth that power corrupts, and that absolute power corrupts absolutely. It was always thus and always will be.

Our absolute power over nature has so thoroughly corrupted us that we are now intent on destroying that part of nature which is our children. Their blood on the pavements – their body parts in the streets. Let us be clear that this is what is coming down the tracks if we fail in our duty. This is what social breakdown looks like – quite simply a living hell for the billions of our young people.

The situation is serious – deadly serious. The propositions in this booklet are serious. This is not an academic exercise. I have been arrested many times and been to prison and I expect the need for this sacrifice to continue.

---

* For more information about Thomas Paine's pamphlet see Wikipedia page 'Thomas Paine (pamphlet)'.

This is where we are now. What is proposed in this booklet is not certain, but based upon the evidence it is our best bet. We can no longer afford the luxury or indulgence of hoping for perfect political and economic conditions. Ignoring the social science on how to bring about rapid and radical political change is as immoral and criminal as ignoring the climate science.

We can delude ourselves if we so wish, but we can be assured the next generation will be free of such delusions. They will be in the middle of the chaos of ecological collapse. They will want the prosecution of those who created the hell they will face. Think on that. Feel it in your body. And then act.

Friends, there are no easy options anymore. There is only one way that leads to true self-respect – and that is Rebellion.

Let's get to it.

References

1. Masson-Delmotte, V., Zhai, P., Pörtner, H.O., Roberts, D., Skea, J., Shukla, P.R., et al (Eds.). *Global warming of 1.5°C. An IPCC Special Report on the impacts of global warming of 1.5°C above pre-industrial levels and related global greenhouse gas emission pathways, in the context of strengthening the global response to the threat of climate change.* World Meteorological Organization Technical Document, sustainable development and efforts to eradicate poverty.

2. Stern, N. (2007) *The Economics of Climate Change: The Stern Review.* Cambridge University Press.

3. Kahan, D. (2010) 'Fixing the communications failure'. *Nature, 463*, 296–297. https://www.nature.com/articles/463296a

4. Masson-Delmotte, V., Zhai, P., Pörtner, H.O., Roberts, D., Skea, J., Shukla, P.R., et al (Eds.) (2018). *Global warming of 1.5°C. An IPCC Special Report on the impacts of global warming of 1.5°C above pre-industrial levels and related global greenhouse gas emission pathways, in the context of strengthening the global response to the threat of climate change.* World Meteorological Organization Technical Document, sustainable development and efforts to eradicate poverty.

5. Xu, Y. & Ramanathan, V. (2017) 'Well below 2°C: Mitigation strategies for avoiding dangerous to catastrophic climate changes'. *Proceedings of the National Academy of Sciences, 114* (39), 10315–10323.

6. Sorokin, P.A. (1925) *The Sociology of Revolution*. J.B. Lippincott & Co.

7. Chenoweth, E. & Stephan, M.J. (2011) *Why Civil Resistance Works: The strategic logic of nonviolent conflict*. Columbia University Press.

8. Lohmann, S. (1994) 'The dynamics of informational cascades: The Monday demonstrations in Leipzig, East Germany, 1989–91'. *World politics, 47*(1), 42–101.

9. Routledge, P. (2010) 'Nineteen days in April: Urban protest and democracy in Nepal'. *Urban Studies, 47*(6) 1279–1299.

10. Chenoweth, E. & Stephan, M.J. (2011) *Why Civil Resistance Works: The strategic logic of nonviolent conflict*. Columbia University Press.

11. Scheidel, W. (2018) *The Great Leveler: Violence and the history of inequality from the Stone Age to the twenty-first century*. Princeton University Press.

12. Skocpol, T. (1979) *States and Social Revolutions: A comparative analysis of France, Russia and China*. Cambridge University Press.

13. Chenoweth, E. & Stephan, M.J. (2011) *Why Civil Resistance Works: The strategic logic of nonviolent conflict*. Columbia University Press.

14. Bastin, J-F., Finegold, Y., Garcia, C., Mollicone, D., Rezende, M. et al (2019) 'The global tree restoration potential'. *Science, 365*(6448), 76–79 https://science.sciencemag.org/content/365/6448/76

---

roger.hallam.uk@gmail.com

http://www.rogerhallam.com

Contact the following organisations to take part in nonviolent direct action:
Extinction Rebellion: https://rebellion.earth
Fridays for Future: https://fridaysforfuture.org
Earth Strike: https://www.earth-strike.com
TransitionLab.earth: https://transitionlab.earth
Transition Network: https://transitionnetwork.org

Or set up your own network or group!